PUBLISHERS' NOTE

As its title implies, the Series in which this volume appears has two purposes. One is to encourage the publication of monographs on advanced or specialised topics in, or related to, the theory and applications of probability and statistics; such works may sometimes be more suited to the present form of publication because the topic may not have reached the stage where a comprehensive treatment is desirable. The second purpose is to make available to a wider public concise courses in the field of probability and statistics, which are sometimes based on unpublished lectures.

The Series was edited from its inception in 1957 by Professor Maurice G. Kendall, under whose editorship the first 21 volumes in the Series appeared. He was succeeded as editor in 1965 by Professor Alan Stuart.

The publishers will be interested in approaches from any authors who have work of importance suitable for the Series.

CHARLES GRIFFIN & CO. LTD.

GRIFFIN'S STATISTICAL MONOGRAPHS AND COURSES

No. 1: *The analysis of multiple time-series* — M. H. QUENOUILLE
No. 2: *A course in multivariate analysis* — M. G. KENDALL
No. 3: *The fundamentals of statistical reasoning* — M. H. QUENOUILLE
No. 4: *Basic ideas of scientific sampling* — A. STUART
No. 6: *An introduction to infinitely many variates* — E. A. ROBINSON
No. 7: *Mathematical methods in the theory of queueing* — A. Y. KHINTCHINE
No. 8: *A course in the geometry of* n *dimensions* — M. G. KENDALL
No. 9: *Random wavelets and cybernetic systems* — E. A. ROBINSON
No. 10: *Geometrical probability* — M. G. KENDALL and P. A. P. MORAN
No. 11: *An introduction to symbolic programming* — P. WEGNER
No. 12: *The method of paired comparisons* — H. A. DAVID
No. 13: *Statistical assessment of the life characteristic: a bibliographic guide* — W. R. BUCKLAND
No. 14: *Applications of characteristic functions* — E. LUKACS and R. G. LAHA
No. 15: *Elements of linear programming with economic applications* — R. C. GEARY and J. E. SPENCER
No. 16: *Inequalities on distribution functions* — H. J. GODWIN
No. 17: *Green's function methods in probability theory* — J. KEILSON
No. 18: *The analysis of variance: a basic course* — A. HUITSON
No. 19: *The linear hypothesis: a general theory* — G. A. F. SEBER
No. 20: *Econometric techniques and problems* — C. E. V. LESER
No. 21: *Stochastically dependent equations: an introductory text for econometricians* — P. R. FISK
No. 22: *Patterns and configurations in finite spaces* — S. VAJDA
No. 23: *The mathematics of experimental design: incomplete block designs and Latin squares* — S. VAJDA
No. 24: *Cumulative sum tests: theory and practice* — C. S. VAN DOBBEN DE BRUYN
No. 25: *Statistical models and their experimental application* — P. OTTESTAD
No. 26: *Statistical tolerance regions: classical and Bayesian* — I. GUTTMAN
No. 27: *Families of bivariate distributions* — K. V. MARDIA
No. 28: *Generalized inverse matrices with applications to statistics* — R. M. PRINGLE and A. A. RAYNER
No. 29: *The generation of random variates* — T. G. NEWMAN and P. L. ODELL
No. 30: *Families of frequency distributions* — J. K. ORD
No. 31: *A first course in statistics* — F. N. DAVID
No. 32: *The logit transformation, with special reference to its uses in bioassay* — W. D. ASHTON
No. 33: *Regression estimation from grouped observations* — Y. HAITOVSKY
No. 34: *Stochastic point processes and their applications* — S. K. SRINIVASAN

No. 5, formerly *Characteristic Functions* by E. Lukacs, is now published independently of the Series

For a list of other statistical and mathematical books see back cover.

THE ANALYSIS OF CATEGORICAL DATA

R. L. PLACKETT, M.A., Sc.D.
Professor of Statistics
University of Newcastle upon Tyne

Monograph No. 35
Series Editor:
ALAN STUART

GRIFFIN LONDON

CHARLES GRIFFIN & COMPANY LIMITED
42 DRURY LANE, LONDON WC2B 5RX

First published 1974
ISBN: 0 85264 228 8

Set by E W C Wilkins Ltd London N12 0EH
Printed in Great Britain by Whitstable Litho Straker Brothers Ltd

CONTENTS

PREFACE

Categorical data arise whenever counts are made instead of measurements. They form an important part of statistical information, especially in medicine and the social sciences. Their analysis has been developed over a long period, the beginning of which is clearly marked by Karl Pearson's famous paper of 1900. Many significant contributions were made by R.A. Fisher. Notwithstanding these early advances, the methodology for counts has been given less attention and has grown more slowly than the methodology for measurements. However, progress has continued steadily, and the subject now has a definite structure which can form the basis of future developments. An earlier negative phase, preoccupied first with the invention of coefficients and later with the formulation of hypotheses, has been replaced by a positive phase in which models are explored. Present conditions therefore justify another addition to the modest number of books which are solely concerned with the analysis of categorical data.

The objects of this book are to describe the principal concepts, explain how they are combined to form statistical methods, illustrate how the methods are applied in practice, indicate the various branches of the subject, and suggest interesting questions which remain to be studied. Readers should have a thorough grasp of the basic principles of statistical inference. The book is intended for students taking advanced courses in statistics, teachers at institutions of higher education, and research workers in methodology or fields of application.

Statistics is concerned with the collection, analysis, and interpretation of data. Consequently the material is presented in the form of a balance between the problems posed by practical examples and the theoretical structure in which they are embedded. Lengthy sections of mathematical argument are avoided by giving proofs in outline or omitting them altogether. A reader who wishes to have further details will usually find that references sufficient for his purpose are placed at suitable points through the text and exercises. Such signposts can be ignored by those who are simply interested in the basic techniques. The main text deals with broad lines of development, and is extended by the exercises, where many subsidiary results are presented as problems. Much the same treatment is accorded to the applications, which involve real data: the text gives interpretations, and the exercises request them.

February, 1974 R.L.P.

ACKNOWLEDGEMENTS

I am grateful to Frank Anscombe, Vic Barnett, Dennis Evans, David Newell and Alan Stuart for their comments and contributions; to Sheila Boyd for the construction of computer programs, much calculation, and drawing the diagrams; to Joyce Edger for preparing the typescript and dealing with correspondence; and to James R. Griffin for helpful advice. I acknowledge the generosity of authors who have allowed me to quote data from their books or papers, and whose names are given in the text. I also wish to thank: Mr J.B. Greenwood; the Trustees of *Biometrika*, the Biometric Society, the Royal Statistical Society, the Massachusetts Medical Society, Messrs Methuen Ltd, and Stanford University Press; the editors of *Annals of Applied Biology*, *British Medical Journal*, *Chemical Engineering Progress*, *Japanese Journal of Geology and Geography*, *Journal of Horticultural Science*, and *Operational Research Quarterly*; the publishers of *Annals of Human Genetics*, *Journal of Experimental Education*, and *Journal of Hygiene*; and the editor and publisher of *British Journal of Preventive and Social Medicine*.

I am solely responsible for any defects that may be found in the book and would be glad to have notification of their occurrence.

1 THE POISSON DISTRIBUTION

1.1 Introduction

A variate N has a Poisson distribution with parameter μ when

$$p(N=j) = \mu^j e^{-\mu}/j! \quad (j = 0, 1, 2, \ldots; \mu > 0).$$

We say that N is $\mathcal{P}(\mu)$. The distribution occupies a central place in the analysis of categorical data. There are several reasons, as follows (Haight, 1967).

(1) The distribution often provides a satisfactory description of experimental data. While many other discrete distributions are available for this purpose, several of the more useful are based on the Poisson distribution.
(2) The distribution can be derived theoretically from elementary principles with a minimum of assumptions, in particular as a limiting form of the binomial and hypergeometric distributions.
(3) If events occur independently and at random in time such that the average rate of occurrence is constant, then the number of events in any fixed time-interval has a Poisson distribution. The same result holds as a limit when many less restricted processes of events in time are superimposed.
(4) The distribution is a member of the exponential family, so that the associated procedures of statistical inference have optimum properties.

These reasons together imply that the Poisson distribution stands in much the same relation to the analysis of categorical data as the normal distribution does in relation to the analysis of measurements made on a continuous scale. The main difference, as we shall see, is that a multiplicative model is appropriate for Poisson means in contrast to a linear model for normal means.

Example 1.1 The data in the first two lines of Table 1.1 are quoted from Drinkwater and Hastings (1967). A record was made of the number of times that each member of a fleet of army vehicles entered the workshop for repair

TABLE 1.1 *Frequency distribution of the number of repair visits for army vehicles*

Number of visits	0	1	2	3	4	5	6+	Total
Frequency observed	295	190	53	5	5	2	0	550
Frequency expected	295·9	183·4	56·8	11·8	1·8	0·2	0·1	550·0

in a year. Here we have events in time, i.e. vehicle failures, which occur much in accordance with the second part of (3), above, since there are many causes of failure for each vehicle, and each cause contributes an individual sequence of failures to the overall sequence. The expected frequencies in the third line of the table are calculated on the assumption that the number of repair visits has a Poisson distribution. They are given by

$$550m^j e^{-m}/j! \quad (j = 0, 1, 2, \ldots),$$

where $m = 0{\cdot}62$ is the average number of repair visits per vehicle. There appears to be good agreement between the observed and expected frequencies. The appropriate methods of inference are explained in Chapter 2.

Example 1.2 A rectangular area in Japan was divided into 1200 equal squares of side 100 metres, and the number of houses in each square was counted. The spatial distribution of the number of houses per square is given by Matui (1932). In Table 1.2, the spatial aspect is ignored. The first two lines give the observed frequency distribution of the number of houses per square, and the third gives the expected frequencies on the assumption of a Poisson distribution. Again, the agreement appears to be satisfactory.

TABLE 1.2 *Frequency distribution of the number of houses in 1200 equal squares of a rectangular area in Japan*

No. houses	0	1	2	3	4	5	6	7	8	9	Total
Freq. obs.	584	398	168	35	9	4	0	1	0	1	1200
Freq. exp.	561·2	426·5	162·1	41·0	7·8	1·2			(0·2)		1200·0

1.2 Properties of Poisson variates

The cumulant-generating function of $\mathcal{P}(u)$ is

$$\log E\{\exp(tN)\} = \mu(e^t - 1),$$

which shows that all the cumulants are equal to μ. In particular,

$$E(N) = \text{var}(N) = \mu.$$

The central moments can be derived from the formulae which connect them with the cumulants (Kendall and Stuart, Vol. 1, ch. 3). Hence, in the usual notation,

$$\mu_3 = \mu \quad \text{and} \quad \mu_4 = \mu + 3\mu^2.$$

Denote a normal distribution with zero mean and unit variance by $\mathcal{N}(0, 1)$. The standardized variate $(N-\mu)/\mu^{\frac{1}{2}}$ converges in distribution to $\mathcal{N}(0, 1)$ as

$\mu \to \infty$. This can be shown as follows. The cumulant-generating function of $(N-\mu)/\mu^{\frac{1}{2}}$ is

$$\mu\{\exp(t\mu^{-\frac{1}{2}})-1\}-t\mu^{\frac{1}{2}} = \tfrac{1}{2}t^2+O(\mu^{-\frac{1}{2}}).$$

In the limit, we obtain the cumulant-generating function of $\mathfrak{N}(0, 1)$. An appeal to the uniqueness theorem for such functions completes the proof.

The distribution function of N can be approximated by

$$p(N \leqslant j) \simeq \Phi\{(j+\tfrac{1}{2}-\mu)/\mu^{\frac{1}{2}}\},$$

where $\Phi(.)$ is the distribution function of $\mathfrak{N}(0, 1)$. The $\frac{1}{2}$ following j is known as a *correction for continuity*, according to the terminology introduced by Yates (1934). An elementary analytical derivation of the correction is outlined by Cox (1970b).

The analysis of Poisson frequencies can be approached by the use of a square-root transformation. In view of the result that

$$\text{var}(N^{\frac{1}{2}}) \to \tfrac{1}{4} \quad \text{as } \mu \to \infty,$$

observations from different Poisson distributions can be transformed so that they have approximately the same variance. The procedures of analysis of variance become applicable provided that a linear model for $E(N^{\frac{1}{2}})$ can be justified. Various modifications of this approach have been put forward (Anscombe, 1948; Freeman and Tukey, 1950), but they are now largely supplanted by procedures devised specially for the purpose and described in the following chapters. We shall find that a linear model for $\log \mu$ is often appropriate. Consider the estimation of $\log \mu$ by $\log(N+c)$, where c is a constant to be determined. The identity

$$\log(N+c) \equiv \log \mu + \log\{1+(N+c-\mu)/\mu\}$$

provides an expansion of $\log(N+c)$ in powers of $(N+c-\mu)/\mu$, whence

$$E\{\log(N+c)\} = \log \mu + (c-\tfrac{1}{2})\mu^{-1}+O(\mu^{-2}).$$

The best value of c is therefore $\frac{1}{2}$, and the estimator $\log(N+\frac{1}{2})$ has a bias which is $O(\mu^{-2})$.

Suppose that $N_1, N_2, \ldots, N_r$ are mutually independent variates such that N_i is distributed as $\mathcal{P}(\mu_i)$ for $i = 1, 2, \ldots, r$. The probability-generating function of ΣN_i is

$$E(s^{\Sigma N_i}) = \Pi E(s^{N_i}) = \exp\{(s-1)\Sigma \mu_i\}.$$

Thus ΣN_i is distributed as $\mathcal{P}(\Sigma \mu_i)$.

In all that follows, the observed value of a variate is denoted by the corresponding lower-case letter. Consider the joint distribution of $\{N_i\}$ conditional on $\Sigma N_i = n$. This is given by

$$p(N_i = n_i \quad \text{for all } i \mid \Sigma N_i = n)$$
$$= \Pi\,(\mu_i^{n_i} e^{-\mu_i}/n_i!)/\{(\Sigma\,\mu_i)^n e^{-\Sigma \mu_i}/n!\}.$$

Denote $\mu_i/\Sigma\,\mu_i$ by p_i. Then $0 < p_i < 1$ and $\Sigma\,p_i = 1$, so that the quantities $\{p_i\}$ form a probability distribution. The conditional joint distribution of $\{N_i\}$ is therefore

$$p(N_i = n_i \quad \text{for all } i \mid \Sigma N_i = n) = n!\,p_1^{n_1} p_2^{n_2} \ldots p_r^{n_r}/n_1!\,n_2! \ldots n_r!,$$

which is a multinomial distribution with index n and probabilities p_1, $p_2, \ldots, p_r$. When $r = 2$, the distribution of N_1 conditional on $N_1 + N_2 = n$ is binomial with index n and probability p_1.

A *mixed* or *compound* Poisson variate K is defined by

$$p(K = k) = \int_0^\infty (x^k e^{-x}/k!)\,f(x)\,dx,$$

where $f(.)$ is the probability density function of a positive variate X. The probability-generating function of K is

$$E(s^k) = \mathfrak{F}(s-1),$$

where $\mathfrak{F}(.)$ is the moment-generating function of X. Consequently

$$E(K) = E(X),$$

and

$$\operatorname{var}(K) = E(X) + \operatorname{var}(X).$$

Thus the variance of a mixed Poisson distribution exceeds the mean.

Example 1.3 X has a gamma distribution with density function

$$f(x) = \theta^\beta x^{\beta-1} e^{-\theta x}/\Gamma(\beta) \quad (x > 0; \theta > 0, \beta > 0).$$

Then K has the negative binomial distribution

$$p(K = k) = \left(\frac{\theta}{1+\theta}\right)^\beta \frac{\Gamma(k+\beta)}{k!\,\Gamma(\beta)} \left(\frac{1}{1+\theta}\right)^k \quad (k = 0, 1, 2, \ldots).$$

According to one hypothesis of accident proneness, industrial workers can be divided into groups such that the number of accidents has a Poisson distribution within groups, and the Poisson parameter has a gamma distribution over the groups. Table 1.3 is taken from Greenwood and Yule (1920), and gives a frequency distribution of the number of accidents to 414 machinists in three months, together with the expected frequencies corresponding to a Poisson distribution and a negative binomial distribution. The parameters of the negative binomial distribution were estimated by equating the observed mean and variance to their theoretical values. An inspection of the table suggests that the negative binomial distribution fits the observed frequencies better than the Poisson distribution.

TABLE 1.3 *Accidents to 414 machinists in three months*

No. accidents	0	1	2	3	4	5+	Total
Frequency observed	296	74	26	8	4	6	414
Expected (Poisson)	256	122	30	5	1	0	414
Expected (negative binomial)	299	69	26	11	5	4	414

Suppose that $Y_1, Y_2, \ldots$ are mutually independent integer-valued variates which are identically distributed and have probability-generating function $g(s)$. A *generalized* Poisson variate Z is defined by

$$Z = Y_1 + Y_2 + \ldots + Y_N,$$

where N is $\mathcal{P}(\mu)$. The probability-generating function of Z is

$$E(s^Z) = \exp\,[\mu\{g(s) - 1\}].$$

Example 1.4 Y has a Poisson distribution with parameter λ. Then Z has a Neyman type A distribution.

Example 1.5 Y has a logarithmic series distribution defined by

$$g(s) = \log(1 - \alpha s)/\log(1 - \alpha) \quad (0 < \alpha < 1).$$

Then Z has a negative binomial distribution.

These and other possibilities are discussed by Skellam (1952) and Feller (1968, ch. XII), who both use the term "compound" in the sense that "generalized" is used here. The analysis of data from generalized Poisson distributions is considered by Hinz and Gurland (1968). A general account is given by Ord (1972).

1.3 Inferences about μ

Suppose that $N_1, N_2, \ldots, N_r$ are mutually independent variates each distributed as $\mathcal{P}(\mu)$. They form a random sample of size r from the distribution. Write $M = \Sigma N_i/r$, so that M is the sample mean. We say that the variate M is an *estimator* of μ, whereas the observed value m is an *estimate*. Given observations $n_1, n_2, \ldots, n_r$, the likelihood is

$$l = \Pi\,(\mu^{n_i}\,e^{-\mu}/n_i!) = \mu^{rm}\,e^{-r\mu}/\Pi\,n_i!$$

The form of l shows that M is sufficient for μ. Furthermore,

$$d \log l/d\mu = r(m - \mu)/\mu.$$

We conclude that M is the unique minimum variance unbiased estimator of μ, with variance μ/r.

Let F_j be the frequency of the event $N = j$. Then

$$\Sigma F_j = r \quad \text{and} \quad \Sigma jF_j = rM.$$

The frequencies $\{F_j\}$ have a multinomial distribution with index r and probabilities $\{p(N = j)\}$ respectively. Consequently,

$$E(F_j) = rp(N = j) = r\mu^j e^{-\mu}/j!$$

This is an *expected frequency* in the strict sense. However, that term is usually attached to the maximum-likelihood estimate of $E(F_j)$, namely $rm^j e^{-m}/j!$, an interpretation which is implicit in section 1.1. The corresponding estimator $rM^j e^{-M}/j!$ is biased, but the unique minimum-variance unbiased estimator can be constructed by using the Rao–Blackwell theorem (Exercise 3, page 115).

A confidence interval for μ is determined as follows. The asymptotic normality of $(N - \mu)/\mu^{\frac{1}{2}}$ implies that $(M - \mu)(r/\mu)^{\frac{1}{2}}$ converges in distribution to $\mathcal{N}(0, 1)$ when r is fixed and $\mu \to \infty$. According to the central limit theorem, $(M - \mu)(r/\mu)^{\frac{1}{2}}$ converges to $\mathcal{N}(0, 1)$ when μ is fixed and $r \to \infty$. Hence, for large μ or large r,

$$p\{(M-\mu)^2(r/\mu) < u^2_{\frac{1}{2}\epsilon}\} \simeq 1 - \epsilon,$$

where $u_{\frac{1}{2}\epsilon}$ is the upper 50ϵ percent point of $\mathcal{N}(0, 1)$. The inequality $(M - \mu)^2 (r/\mu) < \mu^2_{\frac{1}{2}\epsilon}$ is equivalent to $h(\mu) < 0$, where

$$h(\mu) = \mu^2 - \mu(2M + u^2_{\frac{1}{2}\epsilon}/r) + M^2.$$

Consequently the values of μ which satisfy the inequality lie between the roots of $h(\mu) = 0$, which are

$$M + u^2_{\frac{1}{2}\epsilon}/2r \pm u_{\frac{1}{2}\epsilon}(M/r + u^2_{\frac{1}{2}\epsilon}/4r^2)^{\frac{1}{2}}.$$

We conclude that the corresponding observed roots are the limits of a confidence interval for μ which has a confidence coefficient approximately $1 - \epsilon$. When r is large, the limits are essentially $m \pm u_{\frac{1}{2}\epsilon}(m/r)^{\frac{1}{2}}$.

Example 1.6 In Table 1.1, $r = 550$ and $m = 0{\cdot}620$. Take $\epsilon = 0{\cdot}05$. The confidence limits are

$$0{\cdot}620 + 3{\cdot}84/1100 \pm 1{\cdot}96\,(0{\cdot}620/550 + 3{\cdot}84/1\,210\,000)^{\frac{1}{2}}.$$

Thus $(0{\cdot}557, 0{\cdot}689)$ is a confidence interval for μ with a confidence coefficient close to 95 percent.

2 SINGLE CLASSIFICATIONS

2.1 Response

Suppose that the units of a population can be classified into categories on the basis of observable differences. There are three methods by which the categories may be defined, depending on how the differences are expressed.

(i) Purely qualitative differences, e.g. workers classified by occupation, or students classified by combination of subjects.

(ii) A natural ordering, e.g. an attack of influenza classified as mild, normal, or severe.

(iii) Purely quantitative differences, determined by the values of an explanatory variable, e.g. car-drivers classified by the number of accidents in a year, or football players classified by earnings during the last twelve months.

Consider a classification with r categories. A unit selected at random from the population is placed in category i with probability p_i for $i = 1, 2, \ldots, r$. Define a variate R by the probability distribution

$$p(R = i) = p_i \quad (i = 1, 2, \ldots, r).$$

We say that R is a *response*.

Classifications with just two categories are important because they frequently arise in practice. For example, an individual is either male or female, he either smokes or does not smoke, and his weight is either more or less than 68kg (150 lb) exactly. The two categories are conventionally known as *success* and *failure*, where these terms are used in a strictly neutral sense. A classification with two categories defines a *quantal* or *binary* response.

Example 2.1 The data below are taken from Elderton (1920, Table I) and refer to the type of print on the left first finger for fathers.

Arch	Small loop	Large loop	Composite	Whorl
41	139	53	28	68

Here the categories are determined on a purely qualitative basis.

Example 2.2 The data below are quoted by Yates (1948). Children are classified on a graded scale according to the conditions under which homework was carried out. The highest rating is A and the lowest is E.

A	B	C	D	E
308	147	295	179	90

There is a natural ordering among the categories which is reflected in the way they are labelled.

Example 2.3 The data in Table 2.1 are quoted from Lee (1963). All the cases of acute lymphatic leukaemia reported to the British National Cancer Registration Scheme during 1946-60 were classified by month of clinical onset. Here the categories are determined on a quantitative basis.

TABLE 2.1 *Cases of acute lymphatic leukaemia reported to the British Cancer Registration Scheme during 1946-60 and classified by month of clinical onset*

Jan.	Feb.	Mar.	Apr.	May	June	July	Aug.	Sept.	Oct.	Nov.	Dec.
40	34	30	44	39	58	51	55	36	48	33	38

Example 2.4 The data in Table 2.2 are quoted from Edwards and Fraccaro (1960). Swedish families of size 4 or more are classified by the number of boys among the first 4 children. Here again the classification is quantitative.

TABLE 2.2 *Frequencies of boys among the first 4 children in 3343 families of size 4 or more*

Number of boys	0	1	2	3	4
Frequency	183	789	1250	875	246

Example 2.5 The presentation of Table 1.1 is determined by the frequency of repair visits during a year, classified by number, and there are 7 categories. However, the data which underly Table 1.1 consist of the frequency of repair visits classified by vehicle, with 550 categories. The example illustrates how a given set of data may be classified in different ways.

2.2 Homogeneity

Suppose that $N_1, N_2, \ldots, N_r$ are mutually independent Poisson variates with parameters $\mu_1, \mu_2, \ldots, \mu_r$ respectively. Denote by H_0 the hypothesis of homogeneity, according to which

$$\mu_1 = \mu_2 = \ldots = \mu_r.$$

When H_0 holds, the joint distribution of $\{N_i\}$ conditional on $\Sigma N_i = n$ is a symmetric multinomial with $p_i = 1/r$ for $i = 1, 2, \ldots, r$.

There are two principal methods of formulating an alternative hypothesis. According to the first, $N_1, N_2, \ldots, N_r$ are mutually independent Poisson variates where the corresponding parameters $\mu_1, \mu_2, \ldots, \mu_r$ are not all equal. We denote this general alternative by H_1. More restricted versions of H_1 are obtained when $\mu_1, \mu_2, \ldots, \mu_r$ follow a trend or have a periodic fluctuation.

The second type of alternative arises when $N_1, N_2, \ldots, N_r$ are mutually independent variates with the same non-Poisson distribution, in particular a mixed or generalized Poisson. For example, let H_2 be the hypothesis that $\mu_1, \mu_2, \ldots, \mu_r$ are mutually independent gamma variates, each with mean μ and variance σ^2. Then $N_1, N_2, \ldots, N_r$ have a negative binomial distribution with mean μ.

Other possibilities are that $N_1, N_2, \ldots, N_r$ have different distributions or are not independent.

When $r = 2$, the uniformly most powerful unbiased test of the hypothesis $\mu_1 = \mu_2$ against the alternative $\mu_1 \neq \mu_2$, or of $\mu_1 \leqslant \mu_2$ against $\mu_1 > \mu_2$, is obtained by referring N_1 to its distribution conditional on $N_1 + N_2 = n$, supplemented by randomization in order to attain a specified significance level (Lehmann, 1959, pp. 140 ff.). However, randomization is seldom if ever used in practice, mainly for two reasons. First, we are interested more in the level of significance than whether or not a particular value is exceeded; and secondly, inferences should not depend on a device which bears no relation to the problem studied. The conditional distribution of N_1 is binomial with index n and probability $\mu_1/(\mu_1 + \mu_2)$. When $\mu_1 = \mu_2$, the probability is $\frac{1}{2}$, and the variate $(2N_1 - n)/n^{\frac{1}{2}}$ converges to $\mathfrak{N}(0, 1)$ as $n \to \infty$.

Example 2.6 In Table 2.1, the frequency of acute lymphatic leukaemia is 245 for the first six months and 261 for the second six months. Assume that the frequencies are observations on $\mathcal{P}(\mu_1)$ and $\mathcal{P}(\mu_2)$ respectively, and consider the hypothesis $\mu_1 = \mu_2$. The observed value of $(2N_1 - n)/n^{\frac{1}{2}}$ is $-0{\cdot}71$, which gives no evidence for the rejection of $\mu_1 = \mu_2$.

When $r > 2$, no test of homogeneity can be uniformly most powerful against a general alternative. The choice of test criterion will usually lie between the index of dispersion and the likelihood-ratio statistic. Most of what follows is concerned with the first criterion, not because it is preferred but because it has been given more attention. Bayesian tests of homogeneity are discussed by Good (1967).

The *Poisson index of dispersion* X^2 is defined by

$$X^2 = \Sigma (N_i - M)^2/M.$$

X^2 is a special case of Karl Pearson's statistic

$$\Sigma\,(\text{Observed} - \text{Expected})^2/\text{Expected}$$

for testing the agreement of observed and expected frequencies. On this basis, it was proposed as a test of H_0 by Fisher, Thornton and Mackenzie (1922). The procedure of testing H_0 with X^2 is known as the *dispersion* or *variance* test, because $X^2/(r - 1)$ is the ratio of the sample variance to the sample mean. Thus $X^2/(r-1)$ is approximately 1 for a sample from a Poisson distribution but will tend to exceed 1 if $N_1, N_2, \ldots, N_r$ have different Poisson distributions or the same mixed Poisson distribution. These intuitive ideas have been formalized by Potthoff and Whittinghill (1966b), who prove the following result. Among all unbiased tests of H_0 against H_2, the dispersion test is most powerful for σ^2 arbitrarily close to zero, irrespective of the value of μ. Selby (1965) shows that X^2 is a test statistic associated with restricted maximum-likelihood estimation procedures. We shall prove in the following section that X^2 can be derived from the large-sample theory of testing a simple hypothesis against a composite alternative.

When H_0 is true, let μ be the common value of $\mu_1, \mu_2, \ldots, \mu_r$. Since M is sufficient for μ, the distribution of X^2 conditional on $M = m$ is independent of μ, and depends only on n and r. This distribution converges to χ^2_{r-1} as $n \to \infty$ (Cramér, 1946, sect. 30.1). The adequacy of the asymptotic distribution when n is finite has been examined in detail. Consider first the moments of X^2 conditional on $M = m$, which are evaluated by Haldane (1937), Hoel (1943), and Rao and Chakravarti (1956), using various methods. For example,

$$E(X^2 \mid M = m) = (r-1) \quad \text{and} \quad \operatorname{var}(X^2 \mid M = m) = 2(r-1)(1-1/n),$$

as compared with

$$E(\chi^2_{r-1}) = (r-1) \quad \text{and} \quad \operatorname{var}(\chi^2_{r-1}) = 2(r-1).$$

The rapidity with which the third and fourth cumulants tend to those of χ^2_{r-1} depends mainly on m, and partly on r. For small values of n and r, the exact distribution can be enumerated. Tables are given by Rao and Chakravarti (1956), and Chakravarti and Rao (1959). However, the rate at which the number of partitions of n increases with n eventually sets the task of enumeration beyond the power of any computer. Recourse must then be made to Monte Carlo experiments. The results obtained by Cochran (1936), Lancaster and Brown (1965), Slakter (1966), and Good, Gover and Mitchell (1970) show that the agreement between the conditional distribution of X^2 and χ^2_{r-1} can be satisfactory down to quite small values of m.

The observed value of X^2 is computed from the formula

$$r \Sigma n_i^2/n - n,$$

or from the definition of X^2, depending respectively on whether or not a desk-top calculator is used.

Example 2.7 For the data in Table 2.1,

$$r = 12, \quad n = 506, \quad \text{and} \quad \Sigma n_i^2 = 22\,236.$$

The observed value of X^2 is 21·3. Since $\chi^2_{11;0\cdot05} = 19\cdot7$, there is a significant variation in the number of cases from month to month.

Example 2.8 For the data in Table 1.1,

$$r = 550, \quad n = 341, \quad \text{and} \quad \Sigma n_i^2 = 577.$$

When r is large, the significance of X^2 is tested by referring the deviation $(2X^2)^{\frac{1}{2}} - (2r-3)^{\frac{1}{2}}$ to a table of $\mathfrak{N}$ (0, 1). The observed value of X^2 is 589·6, which gives a deviation of 1·217 and implies that a Poisson distribution fits the data satisfactorily.

We now discuss the power of the dispersion test. First, suppose that X^2 is used to test that the distribution of $\{N_i\}$ conditional on $\Sigma N_i = n$ is a symmetric multinomial, against the alternative hypothesis specified by

$$p_i = r^{-1} + c_i n^{-\frac{1}{2}} \quad (i = 1, 2, \ldots, r),$$

where $c_1, c_2, \ldots, c_r$ are fixed and such that $\Sigma c_i = 0$. Write $m_i = np_i$. Then

$$(N_i - m)m^{-\frac{1}{2}} = (N_i - m_i)m_i^{-\frac{1}{2}}(m_i/m)^{\frac{1}{2}} + (m_i - m)m^{-\frac{1}{2}}.$$

Since $m_i/m \to 1$ as $n \to \infty$, we find that X^2 converges in distribution to a non-central χ^2_{r-1} with non-centrality parameter

$$\Sigma (m_i - m)^2/m = r \Sigma c_i^2.$$

This argument is taken from Cochran (1952) and further details are given by Patnaik (1949). An examination of the adequacy of the result when n is finite has been made by numerical comparisons of the exact power of the dispersion test with the nominal power obtained from non-central χ^2_{r-1}. Examples given by Bennett (1959), Haynam and Leone (1965), and Slakter (1968) suggest that the exact power tends to be less than the nominal power in small samples. This is explained, at least in part, by the fact that the exact test is necessarily conservative in the sense that the actual significance level is less than the nominal value.

Secondly, suppose that X^2 is used to test for homogeneity when the alternative hypothesis is that $N_1, N_2, \ldots, N_r$ have the same non-Poisson distribution with mean μ, variance of order μ, and cumulants $\{\kappa_j\}$ which are $O(\mu^{\frac{1}{2}j})$ for $j > 2$. Darwin (1957) proves that $X^2\mu/\kappa_2$ converges in distribution to χ^2_{r-1} as $\mu \to \infty$. The special case of this result for a Neyman type A distribution was first given by Bateman (1950).

We conclude this section by describing the test of H_0 against H_1 based on the likelihood ratio, which is defined by

(Maximum of likelihood under H_0)/(Maximum of likelihood under H_1).

The numerator is $M^{rM}e^{-rM}/\Pi N_i!$ and the denominator is $\Pi (N_i^{N_i}e^{-N_i}/N_i!)$. Consequently the ratio is $\Pi (M/N_i)^{N_i}$. In practice, we use an equivalent test criterion which consists of minus twice the logarithm of the likelihood ratio, and is denoted by Y^2. Hence

$$Y^2 = 2\Sigma N_i \log (N_i/M).$$

According to the general theory of likelihood-ratio tests (Kendall and Stuart, vol. 2, ch. 24), Y^2 converges in distribution to χ^2_{r-1} when H_0 is true. The asymptotic distribution on H_1 is also the same as for X^2. In fact, X^2 and Y^2 are asymptotically equivalent, a result which can be verified by expanding $N_i \log (N_i/M)$ in powers of $(N_i - M)$. Good, Gover and Mitchell (1970) show that the distribution of Y^2 is by no means as well approximated by χ^2_{r-1} as that of X^2 when $m < 1$, and they associate this breakdown with the contributions from $n_i = 0$.

The observed value of Y^2 may be computed from the formula

$$2\Sigma n_i \log n_i - 2n \log m,$$

taking $n_i \log n_i = 0$ when $n_i = 0$.

Example 2.9 The value of Y^2 for the data in Table 2.1 is 21·0, as compared with $X^2 = 21{\cdot}3$.

Example 2.10 The value of Y^2 for the data in Table 1.1 is 793·4, and the deviation $(2Y^2)^{\frac{1}{2}} - (2r-3)^{\frac{1}{2}}$ is 6·71. Thus the hypothesis of a Poisson distribution is rejected. The difference between the conclusions here and in Example 2.8 illustrates the uncertainty attached to the distribution of Y^2 when m is small.

2.3 Large-sample tests

The theory of testing a simple hypothesis against a composite alternative is usually based on the assumption that the n variates observed are independent and identically distributed. However, the results obtained extend immediately to a situation where the variates have neither of these properties. In general terms, suppose that the likelihood L depends on a vector parameter $\boldsymbol{\theta}$, and let $\boldsymbol{\theta}_0$ be a specified value of $\boldsymbol{\theta}$. Denote by $\hat{\boldsymbol{\theta}}$ the maximum-likelihood estimator of $\boldsymbol{\theta}$. Write

$$\kappa_{vw}(\boldsymbol{\theta}) = -n^{-1}E(\partial^2 \log L/\partial\theta_v \partial\theta_w).$$

The convention will be that a function evaluated at $\boldsymbol{\theta} = \hat{\boldsymbol{\theta}}$ has the superscript $\hat{}$, and one evaluated at $\boldsymbol{\theta} = \boldsymbol{\theta}_0$ has the subscript $_0$. Define

$$\mathbf{u} = n^{\frac{1}{2}}(\hat{\boldsymbol{\theta}} - \boldsymbol{\theta}_0), \quad \mathbf{y}_0 = n^{-\frac{1}{2}}[\partial \log L/\partial\theta]_0, \quad \hat{\mathbf{K}} = \{\hat{\kappa}_{vw}\},$$

$$\text{and} \quad \mathbf{K}_0 = \{(\kappa_{vw})_0\}.$$

Three statistics have been proposed to test the hypothesis $\boldsymbol{\theta} = \boldsymbol{\theta}_0$ against the alternative $\boldsymbol{\theta} \neq \boldsymbol{\theta}_0$ in large samples (Rao, 1965, p. 349). They are

$$S_1 = L_0/\hat{L},$$

$$S_2 = \mathbf{u}^T\hat{\mathbf{K}}\mathbf{u},$$

and

$$S_3 = \mathbf{y}_0^T\mathbf{K}_0^{-1}\mathbf{y}_0\,.$$

No single one of these criteria is uniformly best (Peers, 1971). However, S_3 has the practical advantage that the calculation of $\hat{\boldsymbol{\theta}}$ is not required.

We now show how these ideas are relevant to testing the hypothesis of homogeneity, and that S_3 then becomes the Poisson index of dispersion. Define

$$\beta = \log \mu_r \quad \text{and} \quad \theta_a = \log(\mu_a/\mu_r) \quad (a = 1, 2, \ldots, r-1).$$

Then H_0 is equivalent to the simple hypothesis

$$H_\theta : \theta_a = 0 \quad \text{for all } a.$$

The distribution of $\{N_i\}$ is given by

$$\Pi\,\{\exp(n_i \log \mu_i - \mu_i)/n_i!\} = \exp\{-e^\beta(1 + \Sigma\, e^{\theta_a}) + n\beta + \Sigma\, n_a\theta_a\}/\Pi\, n_i!$$

Thus $\Sigma\, N_i$ and $\{N_a\}$ are minimal sufficient for β and $\{\theta_a\}$. We are interested primarily in $\{\theta_a\}$, and can eliminate β by deriving the distribution of $\{N_a\}$ conditional on $\Sigma\, N_i = n$, namely

$$p(N_a = n_a \quad \text{for all } a | \Sigma\, N_i = n) = \{n!\exp(\Sigma\, n_a\theta_a)\}/(1 + \Sigma\, e^{\theta_a})^n\, \Pi\, n_i!$$

This is another version of the multinomial distribution obtained in section 1.2.

We proceed to evaluate the statistic S_3. The elements of $n^{\frac{1}{2}}\mathbf{y}_0$ are typified by

$$(\partial \log L/\partial\theta_a)_0 = N_a - n/r.$$

Similarly, a typical element of $n\mathbf{K}_0$ is

$$-E\,(\partial^2 \log L/\partial\theta_a\,\partial\theta_{a'}) = \mathrm{cov}_0\,(N_a, N_{a'}).$$

The expression $\mathbf{y}_0^T\mathbf{K}_0^{-1}\mathbf{y}_0$ does not involve N_r, and a more symmetrical expression is obtained as follows. Denote by $\mathbf{x}_0$ the vector with elements $\{N_i - n/r\}$, and let $\mathbf{J}_0$ be the matrix with elements $\mathrm{cov}_0\,(N_i, N_{i'})$. $\mathbf{J}_0$ is singular, but there exists a generalized inverse $\mathbf{J}_0^{g_1}$ such that

$$\mathbf{J}_0\mathbf{J}_0^{g_1}\mathbf{J}_0 = \mathbf{J}_0\,.$$

A consequence of the theory of generalized inverses (Pringle and Rayner, 1971) is that

$$\mathbf{y}_0^T\mathbf{K}_0^{-1}\mathbf{y}_0 = \mathbf{x}_0^T\mathbf{J}_0^{g_1}\mathbf{x}_0\,.$$

This result can be proved by the method outlined in Exercise 7, page 116. We therefore have the alternative expression

$$S_3 = \mathbf{x}_0^T \mathbf{J}_0^{g_1} \mathbf{x}_0,$$

which involves all the frequencies $\{N_i\}$.

For a symmetric multinomial distribution,

$$\mathbf{J}_0 = n\mathbf{I}/r - n\mathbf{1}\mathbf{1}^T/r^2,$$

where $\mathbf{I}$ is a unit matrix and $\mathbf{1}$ is a unit vector. The condition $\mathbf{J}_0 \mathbf{J}_0^{g_1} \mathbf{J}_0 = \mathbf{J}_0$ is satisfied by

$$\mathbf{J}_0^{g_1} = r\mathbf{I}/n,$$

and consequently

$$S_3 = \Sigma\,(N_i - m)^2/m.$$

This completes the proof that S_3 is the Poisson index of dispersion. The amount of detail here may appear out of proportion to the results derived. In fact, much of the argument is used again in section 5.3.

2.4 Cycles

Suppose that we have a series of frequencies $N_1, N_2, \ldots, N_r$ which refer to successive equal intervals of time, e.g. hours in a day, or months in a year. The context then implies a particular form of departure from homogeneity, such as a regular oscillation or an isolated peak. This information can be used to construct tests of H_0 which have greater power than the standard tests against unrestricted alternatives. For example, the data in Table 2.1 suggest a summer peak, and the question arises how to incorporate this alternative in the procedure for testing homogeneity.

A model proposed by Fix, Hodges and Lehmann (1959) is

$$\mu_i = \mu + a_i\alpha + b_i\beta,$$

where

$$a_i = 2\sin(\pi/r)\sin\{(2i-1)\pi/r\}, \quad b_i = -2\sin(\pi/r)\cos\{(2i-1)\pi/r\}.$$

The test statistic is

$$T_1 = 2N^{-1}([\Sigma\, N_i \sin\{(2i-1)\pi/r\}]^2 + [\Sigma\, N_i \cos\{(2i-1)\pi/r\}]^2),$$

where N denotes $\Sigma\, N_i$. When the model holds, and $\mu_i \to \infty$ for all i, T_1 converges in distribution to noncentral χ_2^2 with noncentrality parameter $2\,\mu r^2(\alpha^2 + \beta^2)\sin^2(\pi/r)$. Other harmonic terms could be added if appropriate.

Suppose that $r = 2c$, and write

$$U_i = \sum_{j=0}^{c-1} N_{i+j}.$$

The criterion

$$T_2 = N^{-\frac{1}{2}} \max_{1 \leqslant i \leqslant c} |U_i - \tfrac{1}{2}N|$$

was suggested by David and Newell (1965), and they derived an approximation for the percentage points. When their test is applied to the data in Table 2.1, the observed value of T_2 does not quite reach the 1 percent significance level. The power of this test is unknown.

When r is increased indefinitely, the problem becomes that of testing that a set of points on the perimeter of a circle is uniformly distributed. Many criteria have been proposed. The results obtained can be approached through Ajne (1968), who discusses the criterion corresponding to T_2. His work indicates that T_2 is optimal against alternatives where $\mu_i = \xi$ for c successive values of i and $\mu_i = \eta$ for the remainder ($\xi \neq \eta$). The relationship between the problems for finite and infinite values of r suggests other criteria, e.g.

$$T_3 = N^{-1} \sum_{i=1}^{c} (U_i - \tfrac{1}{2}N)^2.$$

Further details are given by Stephens (1969) and Mardia (1972).

The methods of time-series analysis with circular definitions could also be applied to a cycle of frequencies.

2.5 Goodness of fit

Consider a population classified into r categories and let q_j be the probability for category j ($j = 0, 1, \ldots, r-1$). A sample of size f is taken and the frequency of units placed in category j is F_j. We assume that the selection of units corresponds to a sequence of mutually independent events. Under these conditions, the joint distribution of $F_0, F_1, \ldots, F_{r-1}$ is multinomial with index f and probabilities $q_0, q_1, \ldots, q_{r-1}$ respectively. Suppose that these probabilities are specified functions of an unknown parameter θ. For example, the distribution of the number of boys in the first 4 children of families of size 4 or more can be derived from the following two assumptions. First, the probability of a boy is θ for all births; and secondly, successive births form a sequence of mutually independent events. The probability that a family contains j boys is then

$$q_j(\theta) = \binom{4}{j} \theta^j (1-\theta)^{4-j} \quad (j = 0, 1, \ldots, 4).$$

We require to test whether the agreement of the observed frequencies with the statistical model is satisfactory. This is the problem of *goodness of fit*.

The first stage of the analysis consists in the estimation of θ. Given observed frequencies $f_0, f_1, \ldots, f_{r-1}$, the likelihood function is

$$l(\theta; f_0, f_1, \ldots, f_{r-1}) = f! \, \Pi \, [\{q_j(\theta)\}^{f_j}/f_j!],$$

whence

$$\log l = \text{const.} + \Sigma \, f_j \log q_j(\theta).$$

We say that the second term on the right is the *kernel* of the log likelihood: it is the part which depends on θ. Denote by $\hat{\theta}$ the maximum-likelihood estimator of θ. Then $\hat{\theta}$ satisfies the equation

$$\Sigma \, F_j\{\partial \log q_j(\theta)/\partial\theta\} = 0.$$

The maximum-likelihood estimator of $q_j(\theta)$ is $q_j(\hat{\theta})$ for $j = 0, 1, \ldots, r-1$.

In the second stage of the analysis, we test the hypothesis H against the alternative hypothesis K, where H and K are defined as follows. H is the hypothesis that the frequencies $F_0, F_1, \ldots, F_{r-1}$ have a multinomial distribution with index f and probabilities $q_0(\theta), q_1(\theta), \ldots, q_{r-1}(\theta)$ respectively, for some value of θ. K is the hypothesis that the frequencies have a multinomial distribution with index f but without any restriction on the probabilities. As before, there are two test criteria in common use. The first is Pearson's statistic

$$X^2 = \Sigma \, [\{F_j - nq_j(\hat{\theta})\}^2/nq_j(\hat{\theta})].$$

Alternatively, we can use minus twice the logarithm of the likelihood ratio, given by

$$Y^2 = 2 \, \Sigma \, F_j \log \{F_j/nq_j(\hat{\theta})\}.$$

In the limit, as $f \to \infty$, both X^2 and Y^2 converge in distribution to χ^2_{r-2} when H is true. A test of H is therefore made by referring the observed value of either X^2 or Y^2 to a table of χ^2_{r-2} and rejecting H if the value is significantly large.

When $q_j(.)$ is a function of k unknown parameters ($k > 1$), the foregoing procedure is followed with appropriate changes, the chief of which is that the test statistics are now asymptotically χ^2_{r-k-1} when H is true. A rigorous proof of this result for the statistic X^2 is given by Cramér (1946, sect. 30.3). His methods are extended by Mitra (1958) to show that X^2 is asymptotically non-central χ^2_{r-k-1} for a sequence of alternative hypotheses which converges to H as $n \to \infty$.

Example 2.11 Let $f_0, f_1, \ldots, f_4$ be the observed frequencies of 0, 1, ..., 4 boys respectively in the first 4 children of n families with 4 or more children. The likelihood function is

$$l(\theta; f_0, f_1, \ldots, f_4) = f! \, \Pi \left\{ \binom{4}{j} \theta^j (1-\theta)^{4-j} \right\}^{f_j} \Big/ f_j!$$

whence

$$\log l = \text{const.} + \log\theta \, \Sigma \, jf_j + \log(1-\theta) \, \Sigma \, (4-j)f_j.$$

Thus the maximum-likelihood estimator of θ is

$$\hat{\theta} = \Sigma\, jF_j/4n.$$

This is the ratio of the total number of boys to the total number of children. For the data in Table 2.2, we find

$$\hat{\theta} = 0{\cdot}515\,854, \quad X^2 = 0{\cdot}627, \quad Y^2 = 0{\cdot}627.$$

The test criteria are distributed asymptotically as χ_3^2 if the statistical model is correct. We conclude that there is good agreement between the observed frequency distribution and a binomial distribution with index 4 and probability 0·515 854.

Example 2.12 Let $f_0, f_1, f_2, \ldots$ be the observed frequencies of 0, 1, 2, ... respectively in a sample of size f from $\mathcal{P}(\mu)$. The maximum-likelihood estimate of μ is the sample mean

$$m = \Sigma\, jf_j/f.$$

For the data in Table 1.1, $m = 0{\cdot}62$. The expected frequencies are small for 4, 5, and 6+ visits, which could invalidate the χ^2 approximations to X^2 and Y^2. We therefore reclassify the number of visits into 4 categories, namely 0, 1, 2, and 3+. This alters the maximum-likelihood estimate of μ, which now satisfies the equation

$$296 + 12\left(\sum_{j=3}^{\infty} jq_j \Big/ \sum_{j=3}^{\infty} q_j\right) = 550\mu,$$

where

$$q_j = \mu^j e^{-\mu}/j!$$

The solution is 0·6073, and the corresponding expected frequencies are given below, with the values of X^2 and Y^2.

Number of visits	0	1	2	3+	Total
Frequency observed	295	190	53	12	550
Frequency expected	299·65	181·98	55·26	13·11	550·00

$X^2 = 0{\cdot}612 \qquad Y^2 = 0{\cdot}611$

Reference to a table of χ_2^2 indicates that agreement is good.

When the categories of a population are determined by the frequencies of a specific event, as they are in these examples, we say that the statistics X^2 and Y^2 are based on the *frequencies of the frequencies*. In such cases, we may be able to devise a test more powerful against some alternatives where the test statistic is a function of the actual variates instead of the frequencies with which the different values occur. Thus the Poisson index of dispersion is locally more powerful against negative binomial alternatives than are statistics based on the frequencies with which 0, 1, 2, . . . occur. There are similar statistics for testing the homogeneity of a set of binomial or multinomial distributions. The details are given by Potthoff and Whittinghill (1966a) and Wisniewski (1968, 1972).

3 TWO-WAY CLASSIFICATIONS

3.1 Contingency tables

Populations which give rise to the same classification can themselves be classified by qualitative or quantitative differences. For example, consider an experiment designed to study the robustness of a particular make of cup. The experiment consists in dropping cups from a specified height and observing the degree of fracture. A classification of the possible fractures defines the response. Comparable populations are those corresponding to different makes of cup at the same price, different qualities of cup manufactured by the same firm, and different heights for cups of the same make. They exemplify respectively the three methods of expressing differences which are described in section 2.1. A classification of populations defines a *factor*, say $\mathfrak{F}$. By analogy with experiments in agriculture, the jth population in a classified group is known as the jth *level* of $\mathfrak{F}$.

Suppose that a classification with r categories can be made in s populations. A unit selected at random from the jth population is placed in category i with probability p_{ij} for $i = 1, 2, \ldots, r$ and $j = 1, 2, \ldots, s$. The response at the jth level of $\mathfrak{F}$ is a variate R defined by

$$p(R = i;\ \mathfrak{F} = j) = p_{ij} \quad (i = 1, 2, \ldots, r;\ j = 1, 2, \ldots, s).$$

We require to find what differences there are, if any, between the distribution of R at different levels of $\mathfrak{F}$. A sample is taken from each population and the units are classified. Denote by n_{ij} the frequency for the ith category in the jth population. The frequencies can be presented as a rectangular array with r rows and s columns, in which n_{ij} occupies the *cell* formed by the ith row and jth column. This array is known as a *contingency table* of order $r \times s$. For convenience, the table also includes the totals of each row and column, known as the *marginal* totals, together with the total of all frequencies.

Example 3.1 The data in Table 3.1 are quoted from Innes *et al.* (1969). They refer to an experiment designed to test the possible carcinogenic effect of a fungicide, Avadex. Male mice were fed with Avadex at 560 ppm for 85 weeks. Another group of male mice, known as the controls, was kept under standard conditions. The frequency of pulmonary tumours in each group was observed. Here the response is determined by the presence or absence of tumour, and the populations consist of treated mice and controls.

TABLE 3.1 *Numbers of mice bearing tumours in treated and control groups*

Condition of mice	Treated	Control	Total
Tumours	4	5	9
No tumours	12	74	86
Total	16	79	95

Thus, both factor and response are defined by qualitative differences. The proportion of tumours is higher for the treated group and we need to assess the significance of this observation.

Example 3.2 The data in Table 3.2 are quoted from Hewlett and Plackett (1950). They are concerned with the toxicity to the beetle *Tribolium castaneum* of films formed by the insecticide γ-benzene hexachloride. The populations are defined by the deposit of 0·1% w/v γ-BHC measured in

TABLE 3.2 *Toxicity to Tribolium castaneum of films formed by 0·1%* w/v *γ-benzene hexachloride*

Period of survival (days)	Deposit of γ-BHC ($mg/10\ cm^2$)						Total
	12·08	14·49	16·31	18·13	20·44	22·26	
At most 6	15	24	26	24	29	29	147
Between 7 and 9	5	4	7	6	4	4	30
More than 9	30	21	17	20	17	16	121
Total	50	49	50	50	50	49	298

$mg/10\ cm^2$. Response is classified according to the period of survival following the application of the insecticide. The categories are: death within six days, death between seven and nine days, and survival for more than nine days. Thus both factor and response are defined by quantitative differences. We wish to determine the relationship between the proportions in the various categories and the dose of γ-BHC.

Suppose next that the units of a population can be divided into categories by each of two methods of classification, which contain r and s categories respectively. A unit selected at random from the population is placed with probability p_{ij} in category i of the first classification and category j of the second. Define variates R and S by the probability distribution

$$p(R=i, S=j) = p_{ij} \quad (i=1, 2, \ldots, r;\ j=1, 2, \ldots, s).$$

Thus the double classification defines a bivariate response (R, S). We are now concerned with the problem of describing the distribution of R and S. A sample is taken from the population and the units are classified with respect to each response. Denote by n_{ij} the frequency for the ith category of the first classification and the jth category of the second. The frequencies $\{n_{ij}\}$ can again be presented as a contingency table of order $r \times s$.

Example 3.3 The data in Table 3.3 are quoted by Armitage (1955) from Holmes and Williams (1954) with further details. They show the relationship

TABLE 3.3 *Relationship between nasal carrier rate for Streptococcus pyogenes and size of tonsils among 1398 children aged 0–15 years*

	Present, but not enlarged	Enlarged tonsils		Total
	+	++	+++	
Carriers	19	29	24	72
Non-carriers	497	560	269	1326
Total	516	589	293	1398
Carrier-rate	0·0368	0·0492	0·0819	

between nasal carrier rate for *Streptococcus pyogenes* and size of tonsils among 1398 children aged 0–15 years. Children are classified qualitatively as carriers and non-carriers, whereas there is a natural ordering for the size of tonsils. The carrier rate increases with tonsil size and suggests a particular form for the joint distribution of responses.

Example 3.4 The data in Table 3.4 are quoted by Yates (1948). They were obtained in the course of a pilot enquiry into the conditions in which

TABLE 3.4 *Classification of 1019 children according to conditions under which homework was carried out, and the teacher's rating of the quality of that homework. (Each scale is graded, A being the highest rating)*

Teacher's rating	Homework conditions					Total
	A	*B*	*C*	*D*	*E*	
A	141	67	114	79	39	440
B	131	66	143	72	35	447
C	36	14	38	28	16	132
Total	308	147	295	179	90	1019

schoolchildren do their homework. Children are classified according to the conditions under which homework was carried out, and according to the teacher's rating of the quality of that homework. Although the effect of the homework conditions on the quality of the preparation appears to be small, the question arises whether the slight trend observed has any significance.

Example 3.5 The data in Table 3.5 are quoted by Stuart (1953). They are based on case-records of the eye-testing of women employees in Royal

TABLE 3.5 *Unaided distance vision of 7477 women aged 30–39*

Grade of right eye	Grade of left eye				Total
	Highest	Second	Third	Lowest	
Highest	1520	266	124	66	1976
Second	234	1512	432	78	2256
Third	117	362	1772	205	2456
Lowest	36	82	179	492	789
Total	1907	2222	2507	841	7477

Ordnance factories during 1943–6. There are two responses, defined by vision in the right and left eyes. The categories are the same for each response, and ordered. A hypothesis of possible interest is that the margins of the table are homogeneous, in the sense that each response has the same probability distribution.

Example 3.6 The data in Table 3.6 are quoted by Waite (1915). Fingerprints of the right hand are classified by the numbers of whorls and small

TABLE 3.6 *Fingerprints of the right hand classified by the number of whorls and small loops*

Whorls	Small loops						Total
	0	1	2	3	4	5	
0	78	144	204	211	179	45	861
1	106	153	126	80	32		497
2	130	92	55	15			292
3	125	38	7				170
4	104	26					130
5	50						50
Total	593	453	392	306	211	45	2000

loops. Each fingerprint is either whorl, arch, composite, or one of four varieties of loop. Consequently the total of whorls and small loops for each hand is at most 5, and the table is triangular. The question arises of constructing an appropriate model for the corresponding probabilities.

Although the concepts of response and factor have a logical appeal and are widely used, they are often difficult to distinguish in practice. There are few experiments where the sample sizes are previously determined. Thus the column totals in Table 3.1, although treated as fixed, have every appearance of being fortuitous. The descriptions of experimental procedures given in papers mainly concerned with statistical methodology may permit various allocations of the labels "response" or "factor". A possible solution of the difficulty is to identify factors and responses with causes and effects respectively, but an inspection of sets of data suggests that such identification can also be ambiguous. We shall continue for the present to suppose that responses and factors can be clearly separated, because this serves to introduce and describe other relevant concepts. In a sense, the distinction is unimportant. The basic statistical model is one where the frequencies are assigned independent Poisson distributions. Under this model, constraints on the frequencies are equivalent to the definition of factors. We shall find that the analysis of data is largely independent of such constraints. Consequently, a failure to distinguish fully between responses and factors serves only to modify the presentation of the results.

3.2 Variation of response

A response R is observed at each level of a factor $\mathfrak{F}$, where R is defined by a classification with r categories, and $\mathfrak{F}$ by a set of s populations. The distribution of R at the jth level of $\mathfrak{F}$ is expressed by

$$p(R = i; \mathfrak{F} = j) = p_{ij} \quad (i = 1, 2, \ldots, r;\ j = 1, 2, \ldots, s),$$

where

$$p_{1j} + p_{2j} + \ldots + p_{rj} = 1 \quad \text{for all } j.$$

We assume that $p_{ij} > 0$ for all i and j. Define

$$\theta_a = \log(p_{as}/p_{rs}) \quad \text{and} \quad \lambda_{ab} = \log(p_{ab}p_{rs}/p_{as}p_{rb}),$$

where $a = 1, 2, \ldots, r-1$, and $b = 1, 2, \ldots, s-1$. Consider the hypothesis

$$H_1 : p_{i1} = p_{i2} = \ldots = p_{is} \quad \text{for } i = 1, 2, \ldots, r.$$

This specifies that the distribution of R is the same at each level of $\mathfrak{F}$. Then H_1 is equivalent to the hypothesis

$$H_\lambda : \lambda_{ab} = 0 \quad \text{for all } a \text{ and } b.$$

The result is proved by showing that each hypothesis implies the other. If H_1 is true, then so is H_λ. Conversely, suppose that H_λ holds and consider the matrix $\mathbf{P} = \{p_{ij}\}$. Since

$$p_{ab}p_{rs} = p_{as}p_{rb} \quad \text{for all } a \text{ and } b,$$

we deduce that $\mathbf{P}$ has rank 1. Thus vectors $\mathbf{u}$ and $\mathbf{v}$ exist such that $\mathbf{P} = \mathbf{u}\mathbf{v}^T$. However, $\mathbf{1}^T\mathbf{P} = \mathbf{1}^T$, where $\mathbf{1}$ is a vector of units, whence $\mathbf{v} = \mathbf{1}/\mathbf{1}^T\mathbf{u}$ and $\mathbf{P} = \mathbf{u}\mathbf{1}^T/\mathbf{1}^T\mathbf{u}$. This shows that H_1 holds, and completes the proof that H_1 and H_λ are equivalent. When H_λ holds, we can write

$$p_{ij} = p_i \quad \text{for all } i \text{ and } j.$$

The hypothesis

$$H_\theta : \theta_a = 0 \quad \text{for all } a$$

then ensures that

$$p_{1j} = p_{2j} = \ldots = p_{rj} = 1/r \quad \text{for all } j.$$

Example 3.7 When $r = s = 2$, the probabilities $\{p_{ij}\}$ are such that

$$p_{11} + p_{21} = 1 \quad \text{and} \quad p_{12} + p_{22} = 1.$$

We say that $p_{11}p_{22}/p_{12}p_{21}$ is the *odds ratio* because of the alternative expression

$$\{p_{11}/(1 - p_{11})\}/\{p_{12}/(1 - p_{12})\}.$$

This quantity is also known as the *relative risk*, a term more accurately applied to p_{11}/p_{12}. There is just one parameter in the class $\{\lambda_{ab}\}$, which we denote by λ. Thus

$$\lambda = \log\{p_{11}/(1 - p_{11})\} - \log\{p_{12}/(1 - p_{12})\},$$

which shows that $p_{11} - p_{12}$ and λ are negative, zero and positive together. Consequently, hypotheses about the sign of $p_{11} - p_{12}$ are equivalent to hypotheses about the sign of λ. More generally, inferences about $p_{11} - p_{12}$ are conveniently replaced by inferences about λ. This corresponds to measuring probabilities on a logit scale, where $\log\{p/(1 - p)\}$ is known as the *logit* of the probability p.

A sample is taken from each population, and n_{ij} is the frequency observed in the ith category for the jth population. We write

$$n_{io} = \Sigma_j n_{ij}, \quad n_{oj} = \Sigma_i n_{ij}, \quad \text{and} \quad n = \Sigma_{ij} n_{ij},$$

where Σ_i denotes a summation over the possible values of i, and Σ_j and Σ_{ij} are similarly defined. Let N_{ij} be the variate corresponding to n_{ij}. The variates N_{io} and N_{oj} are defined by analogy with n_{io} and n_{oj} respectively. Since the marginal total N_{oj} is fixed at n_{oj} for all j, the contingency table takes the form shown in Table 3.7.

TABLE 3.7 *Contingency table for one response and one factor*

Response categories	Populations 1	2	...	s	Total
1	N_{11}	N_{12}	...	N_{1s}	N_{10}
2	N_{21}	N_{22}	...	N_{2s}	N_{20}
⋮	...	...	...	...	⋮
r	N_{r1}	N_{r2}	...	N_{rs}	N_{r0}
Total	n_{01}	n_{02}	...	n_{0s}	n

We assume that units are taken independently and at random, whence two consequences follow:

(a) The vector $\mathbf{N}_j = (N_{1j}, N_{2j}, \ldots, N_{rj})^T$ has a multinomial distribution with index n_{0j} and probabilities $(p_{1j}, p_{2j}, \ldots, p_{rj})$.

(b) Such vectors are mutually independent for $j = 1, 2, \ldots, s$. The distribution of $\{N_{ij}\}$ is therefore given by

$$p(N_{ij} = n_{ij} \quad \text{for all } i \text{ and } j) = \Pi_{ij}\, n_{0j}!p_{ij}^{n_{ij}}/n_{ij}!,$$

where Π_{ij} is a product over the possible values of i and j. This distribution can be derived as follows from rs mutually independent variates $\{N_{ij}\}$ such that N_{ij} has a Poisson distribution with parameter μ_{ij}. Denote $\Sigma_i\, \mu_{ij}$ by μ_{0j}. When a classification with r categories is made in s populations, the conditions $\{N_{0j} = n_{0j}\}$ are imposed. The distribution of $\{N_{ij}\}$ conditional on $\{N_{0j} = n_{0j}\}$ consists of s mutually independent multinomial distributions, the jth of which has index n_{0j} and probabilities $\{p_{ij}\}$, where $p_{ij} = \mu_{ij}/\mu_{0j}$.

In terms of the parameters $\{\theta_a\}$ and $\{\lambda_{ab}\}$, the distribution of $\{N_{ij}\}$ is given by

$$p(N_{ij} = n_{ij} \quad \text{for all } i \text{ and } j) =$$
$$\exp\,(\Sigma_a\, n_{a0}\theta_a + \Sigma_{ab}\, n_{ab}\lambda_{ab})\, \Pi_{ij}\, (n_{0j}!p_{rj}^{n_{0j}}/n_{ij}!).$$

The equivalence of this expression to the one given previously is verified by checking that the powers of p_{ab}, p_{as}, p_{rb} and p_{rs} agree. For example, the coefficient of log p_{as} in the exponential term is

$$n_{a0} - \Sigma_b\, n_{ab} = n_{as}.$$

The statistics $\{N_{a0}\}$ and $\{N_{ab}\}$ are minimal sufficient for $\{\theta_a\}$ and $\{\lambda_{ab}\}$. We are primarily interested in $\{\lambda_{ab}\}$ and therefore consider the distribution of $\{N_{ab}\}$ conditional on $\{N_{a0} = n_{a0}\}$. The fact that the column totals are fixed provides the further condition $\{N_{0j} = n_{0j}\}$. Thus the distribution of $\{N_{a0}\}$ is

$$p(N_{a0} = n_{a0} \quad \text{for all } a) = \Sigma_{n_{ab}}\, p(N_{ij} = n_{ij} \quad \text{for all } i \text{ and } j),$$

where the summation is taken over all $\{n_{ab}\}$ which satisfy both the conditions. Hence the conditional distribution required is

$$p(N_{ab} = n_{ab} | N_{a0} = n_{a0} \quad \text{for all } a \text{ and } b)$$

$$= \{\exp(\Sigma_{ab}\, n_{ab}\lambda_{ab})/\Pi_{ij}\, n_{ij}!\}/\{\Sigma_{n_{ab}} \exp(\Sigma_{ab}\, n_{ab}\lambda_{ab})/\Pi_{ij}\, n_{ij}!\}.$$

We denote the expression on the right by $h\{n_{ab}; n_{i0}, n_{0j}, \lambda_{ab}\}$. This is also the distribution of $\{N_{ij}\}$ conditional on $\{N_{i0} = n_{i0}\}$ and $\{N_{0j} = n_{0j}\}$, when the $\{N_{ij}\}$ are mutually independent Poisson variates with parameters $\{\mu_{ij}\}$ respectively, and

$$\lambda_{ab} = \log(\mu_{ab}\mu_{rs}/\mu_{as}\mu_{rb}).$$

Example 3.8 When $r = s = 2$, the conditions on $\{n_{ij}\}$ imply that

$$n_{12} = n_{10} - n_{11}, \quad n_{21} = n_{01} - n_{11},$$

and

$$n_{22} = n_{11} - (n_{10} + n_{01} - n).$$

Hence the possible values of n_{11} satisfy

$$v \leqslant n_{11} \leqslant w,$$

where $v = \max(0, n_{10} + n_{01} - n)$ and $w = \min(n_{10}, n_{01})$.

The conditional distribution of N_{11} is

$$h(n_{11}; n_{i0}, n_{0j}, \lambda) = \{\exp(n_{11}\lambda)/\Pi_{ij}\, n_{ij}!\} \Big/ \left\{\sum_{n_{11}=v}^{w} \exp(n_{11}\lambda)/\Pi_{ij}\, n_{ij}!\right\}.$$

This distribution was first derived by Fisher (1935).

When H_λ holds, the distribution of R is the same for each level of $\mathfrak{F}$, and $p_{ij} = p_i$ for all i and j. The distribution of $\{N_{a0}\}$ is the convolution of s multinomial distributions with indices $n_{01}, n_{02}, \ldots, n_{0s}$ and the same probabilities $\{p_i\}$. Hence $\{N_{a0}\}$ has a multinomial distribution with index n and probabilities $\{p_i\}$. Inferences about the probabilities are made along the lines discussed in Chapter 2.

The main part of the analysis is therefore concerned with inferences about $\{\lambda_{ab}\}$ from the distribution with probabilities $h\{n_{ab}; n_{i0}, n_{0j}, \lambda_{ab}\}$. We return to this topic in the following two chapters.

3.3 Association between responses

A population is classified by each of two methods, which contain r and s categories respectively. The corresponding bivariate response is (R, S), where

$$p(R = i, S = j) = p_{ij} \quad (i = 1, 2, \ldots, r;\ j = 1, 2, \ldots, s),$$

and

$$\Sigma_{ij}\, p_{ij} = 1.$$

We assume as before that $p_{ij} > 0$ for all i and j. This assumption is not fulfilled for Example 3.6. The relevant issues are discussed in section 5.6, below.

Consider the hypothesis

$$H_2 : p_{ij} = p_{io}p_{oj} \quad \text{for all } i \text{ and } j,$$

where $p_{io} = \Sigma_j \, p_{ij}$ and $p_{oj} = \Sigma_i \, p_{ij}$. This specifies that the responses are independent, so that $p(R = i \mid S = j)$ is the same for all j; and similarly with the responses interchanged. When the responses are not independent, there is an *association* between them, and the question arises of how the association can be measured. Consider a 2 × 2 table in which $R = 1$ and $S = 1$ are the categories of primary interest for the two binary responses. Then these categories are positively associated if either of the following equivalent conditions holds:

(i) $p_{11} > p_{10}p_{01}$,

and (ii) $p(R = 1 \mid S = 1) > p(R = 1 \mid S = 2)$.

When the inequalities are reversed, the categories $R = 1$ and $S = 1$ are *negatively associated*.

Example 3.9 Table 3.3, above, shows the relationship between nasal carrier rate for *Streptococcus pyogenes* and size of tonsils. By combining the categories of enlarged tonsils, we obtain the 2 × 2 table given below.

Tonsils	Carriers	Non-carriers
Enlarged	53	829
Not enlarged	19	497

The proportion with enlarged tonsils among carriers is greater than among non-carriers. There appears to be a positive association between the categories of carrier and enlarged tonsils.

Many coefficients of association for 2 × 2 tables have been proposed, and they are reviewed by Kendall and Stuart (vol. 2, ch. 33). Such coefficients take the value zero when the responses are independent, and are positive or negative depending on the type of association. They are purely descriptive and have no real interpretation. Considerations of symmetry (Simpson, 1951; Edwards, 1963) indicate that the measure of association for a 2 × 2 table should be some function of

$$\psi = p_{11}p_{22}/p_{12}p_{21},$$

which is known as the *cross-product ratio* of the table. A convenient function is

$$\lambda = \log(p_{11}p_{22}/p_{12}p_{21}).$$

Thus R and S are independent when $\lambda = 0$, while the two types of association correspond respectively to $\lambda > 0$ and $\lambda < 0$.

We now turn to $r \times s$ tables. The absence of motivation in early coefficients led Goodman and Kruskal (1954, 1959, 1963, 1972) to introduce coefficients for $r \times s$ tables with probabilistic interpretations. They have also derived the tests appropriate for various sampling procedures. On the other hand, the symmetry argument for 2×2 tables can be extended to $r \times s$ tables (Altham, 1970a), and shows that association ought to be measured by some function of the $(r-1)(s-1)$ cross-product ratios

$$\psi_{ab} = p_{ab}p_{rs}/p_{as}p_{rb} \quad (a = 1, 2, \ldots, r-1;\ b = 1, 2, \ldots, s-1).$$

Equivalently, the measure of association is a function of the parameters

$$\lambda_{ab} = \log \psi_{ab}.$$

Further reasons for the importance of $\{\lambda_{ab}\}$ will appear in the following section.

An analysis of the association between R and S is based on a transformation of the probabilities $\{p_{ij}\}$ to new parameters, which consist of $\{\lambda_{ab}\}$ and

$$\kappa = \log p_{rs}, \quad \theta_a = \log(p_{as}/p_{rs}), \quad \text{and} \quad \phi_b = \log(p_{rb}/p_{rs}).$$

The hypothesis of independence H_2 is equivalent to

$$H_\lambda : \lambda_{ab} = 0 \quad \text{for all } a \text{ and } b.$$

The proof follows the same lines as in section 3.2. If H_2 is true, then so is H_λ. Conversely, H_λ implies that $\mathbf{P} = \mathbf{uv}^T$. The condition that $\mathbf{1}^T\mathbf{P1} = 1$ ensures that $\mathbf{P} = (\mathbf{P1})(\mathbf{P}^T\mathbf{1})^T$, which is H_2.

Suppose that H_λ is true. Then the hypothesis

$$H_\phi : \phi_b = 0 \quad \text{for all } b$$

implies that the columns of $\mathbf{P}$ are identical. When H_λ and H_ϕ both hold, the hypothesis

$$H_\theta : \theta_a = 0 \quad \text{for all } a$$

ensures that each element of $\mathbf{P}$ is $1/rs$. The same conclusion follows from H_ϕ when H_λ and H_θ both hold.

There is a similar procedure of considering a sequence of hypotheses in the analysis of variance of a two-way layout. The comparison suggests that the parameters $\{\theta_a\}$, $\{\phi_b\}$, and $\{\lambda_{ab}\}$ can be described as the main effect of R, the main effect of S, and the *first-order* or *two-response interaction*

$R \times S$ respectively. In such terms, H_λ is the hypothesis of zero first-order interaction.

A sample of size n is taken from the population, and n_{ij} is the frequency observed in the ith category of the first classification and the jth category of the second. The resulting contingency table takes the form shown in Table 3.8, where again N_{ij} is the variate corresponding to n_{ij}. We assume

TABLE 3.8 *Contingency table for two responses*

Categories of first response	Categories of second response 1	2	....	s	Total
1	N_{11}	N_{12}	....	N_{1s}	N_{10}
2	N_{21}	N_{22}	...	N_{2s}	N_{20}
⋮	...	...	...	...	⋮
r	N_{r1}	N_{r2}	...	N_{rs}	N_{r0}
Total	N_{01}	N_{02}	...	N_{0s}	n

that units are taken independently and at random, in which case the variates $\{N_{ij}\}$ have a multinomial distribution with index n and probabilities $\{p_{ij}\}$. The distribution of $\{N_{ij}\}$ is thus

$$p(N_{ij} = n_{ij} \quad \text{for all } i \text{ and } j) \;=\; n!\, \Pi_{ij}(p_{ij}^{n_{ij}}/n_{ij}!).$$

We can derive the same result from rs mutually independent variates $\{N_{ij}\}$ such that N_{ij} has a Poisson distribution with mean μ_{ij}. Denote $\Sigma_{ij}\, \mu_{ij}$ by μ. When a population is classified by each of two methods and a sample of size n taken, the condition $\Sigma_{ij}\, N_{ij} = n$ is imposed. The distribution of $\{N_{ij}\}$ conditional on $\Sigma_{ij}\, N_{ij} = n$ is multinomial with index n and probabilities $\{p_{ij}\}$, where $p_{ij} = \mu_{ij}/\mu$.

In terms of the new parameters, the distribution of $\{N_{ij}\}$ is given by

$$p(N_{ij} = n_{ij} \quad \text{for all } i \text{ and } j)$$
$$= \exp(n\kappa + \Sigma_a\, n_{a0}\theta_a + \Sigma_b\, n_{0b}\phi_b + \Sigma_{ab}\, n_{ab}\lambda_{ab})n!/\Pi_{ij}n_{ij}!$$

Thus $\{N_{a0}\}$, $\{N_{0b}\}$, and $\{N_{ab}\}$ are minimal sufficient for $\{\theta_a\}$, $\{\phi_b\}$, and $\{\lambda_{ab}\}$. We are primarily interested in $\{\lambda_{ab}\}$, and therefore derive the distribution of $\{N_{ab}\}$ conditional on $\{N_{a0} = n_{a0}\}$ and $\{N_{0b} = n_{0b}\}$. By the same arguments as in section 3.2, the conditional distribution required is

$$p(N_{ab} = n_{ab} | N_{a0} = n_{a0} \text{ and } N_{0b} = n_{0b} \text{ for all } a \text{ and } b) = h\{n_{ab}; n_{i0}, n_{0j}, \lambda_{ab}\}.$$

When H_λ holds, the responses R and S are independent, in which case $\{N_{a0}\}$ and $\{N_{0b}\}$ have independent multinomial distributions with index n

and probabilities $\{p_{io}\}$ and $\{p_{oj}\}$ respectively. We have described in Chapter 2 how to make inferences from multinomial distributions. Thus the main part of the analysis is again concerned with inferences about $\{\lambda_{ab}\}$ from the distribution with probabilities $h\{n_{ab}; n_{io}, n_{oj}, \lambda_{ab}\}$. We return to this topic in the following two chapters.

3.4 Contingency tables with given margins

Suppose that the distribution of a bivariate response (R, S) is defined by

$$p(R=i, S=j) = p_{ij} \quad (i=1,2,\ldots,r;\ j=1,2,\ldots,s).$$

In the previous section we introduced a transformation from the probabilities $\{p_{ij}\}$ to parameters κ, $\{\theta_a\}$, $\{\phi_b\}$, and $\{\lambda_{ab}\}$. The reverse transformation is given by

$$\log p_{rs} = \kappa, \quad \log p_{as} = \kappa + \theta_a, \quad \log p_{rb} = \kappa + \phi_b,$$

and

$$\log p_{ab} = \kappa + \theta_a + \phi_b + \lambda_{ab} \quad (a=1,2,\ldots,r-1;\ b=1,2,\ldots,s-1).$$

These equations constitute a linear model for the logarithms of the probabilities $\{p_{ij}\}$, known as a *log-linear* model. The equations can be solved uniquely for the new parameters, and so the model is described as *saturated.* A hypothesis, as distinct from a tautology, is expressed only when some of the parameters have specified values, in which case the model is *unsaturated.* There is an analogy with the analysis of variance, where main effects and interactions are defined in terms of the expected values of the observations, with the consequence that the expected values can be expressed as linear combinations of factorial effects. An alternative version of the log-linear model is suggested by the standard analysis of variance model for cross-classified factorial designs, and was introduced by Birch (1963). According to the second version, we write

$$\log p_{ij} = \mu + \alpha_i + \beta_j + \delta_{ij},$$

where $$\Sigma_i \alpha_i = \Sigma_j \beta_j = \Sigma_i \delta_{ij} = \Sigma_j \delta_{ij} = 0.$$

The two models are different, but are related when $r = s = 2$, in which case $4\delta_{11} = \lambda_{11}$. We shall continue to use the first version, which was proposed in general terms by Mantel (1966).

The probabilities $\{p_{ij}\}$ are uniquely determined by the combination of parameters κ, $\{\theta_a\}$, $\{\phi_b\}$ and $\{\lambda_{ab}\}$. We proceed to show that the $\{p_{ij}\}$ are also uniquely determined by a combination which consists of the parameters $\{\lambda_{ab}\}$ and the marginal probabilities $\{p_{io}\}$ and $\{p_{oj}\}$. This result confirms the basic positions of $\{\lambda_{ab}\}$ in the measurement of association and will also be found important in the asymptotic behaviour of the distribution $h\{n_{ab}; n_{io}, n_{oj}, \lambda_{ab}\}$.

Consider first the case $r = s = 2$. Since $\{p_{io}\}$ and $\{p_{oj}\}$ are given, the probabilities are expressed in terms of p_{11} by

$$p_{12} = p_{10} - p_{11}, \quad p_{21} = p_{01} - p_{11}, \quad \text{and} \quad p_{22} = p_{11} - p_{10} - p_{01} + 1.$$

Hence p_{11} is subject to the conditions

$$\max(0, p_{10} + p_{01} - 1) \leqslant p_{11} \leqslant \min(p_{10}, p_{01}).$$

The cross-product ratio ψ is also given, whence

$$p_{11}(p_{11} - p_{10} - p_{01} + 1)/(p_{10} - p_{11})(p_{01} - p_{11}) = \psi.$$

Thus p_{11} satisfies the quadratic equation

$$p_{11}^2(\psi - 1) - p_{11}\{(\psi - 1)(p_{10} + p_{01}) + 1\} + \psi p_{10} p_{01} = 0.$$

The left-hand side is positive at the lower limit of p_{11} and negative at the upper limit. We conclude that only one root of the quadratic equation lies in the permissible range of p_{11}. Therefore the probabilities $\{p_{ij}\}$ are uniquely determined by the combination of p_{10}, p_{01}, and λ.

The generalization of this result to a table of order $r \times s$ is based on the following theorem, which is due to Sinkhorn (1967). Let $\{p_{io}\}$ and $\{p_{oj}\}$ be specified marginal probabilities. Then, corresponding to each positive matrix **Q** of order $r \times s$, there is a unique matrix **CQD** with row sums $\{p_{io}\}$ and column sums $\{p_{oj}\}$, where **C** and **D** are respectively $r \times r$ and $s \times s$ diagonal matrices with positive diagonals and are themselves unique up to a scalar multiple. Furthermore, **CQD** can be found by a convergent iterative process which consists of alternately scaling the rows and columns of **Q** to have row sums $\{p_{io}\}$ and column sums $\{p_{oj}\}$ respectively. A proof of the theorem is omitted.

Suppose that we are given the marginal probabilities $\{p_{io}\}$ and $\{p_{oj}\}$ of a table of order $r \times s$, together with the parameters $\{\lambda_{ab}\}$. Thus the probabilities $\{p_{ij}\}$ satisfy the constraints

$$\Sigma_j\, p_{ij} = p_{io}, \quad \Sigma_i\, p_{ij} = p_{oj}, \quad \text{and} \quad \log(p_{ab}p_{rs}/p_{as}p_{rb}) = \lambda_{ab}.$$

We prove as follows that there is a unique solution. Denote the cross-product ratio $\exp(\lambda_{ab})$ by ψ_{ab}. Let **Q** be any positive matrix which has order $r \times s$ and cross-product ratios $\{\psi_{ab}\}$. We may construct **Q** by assigning arbitrary positive values to its last row and column, and then calculating the remaining elements from

$$q_{ab} = \psi_{ab} q_{as} q_{rb} / q_{rs}.$$

The choice $q_{as} = q_{rb} = q_{rs} = 1$ is usually the most convenient. According to the theorem stated above, there is a unique matrix **CQD** with the specified marginal totals. Moreover, the cross-product ratios of **CQD** are given by

$$(q_{ab}c_ad_b)(q_{rs}c_rd_s)/(q_{as}c_ad_s)(q_{rb}c_rd_b) = \psi_{ab}.$$

Thus **CQD** satisfies all the conditions stated and is the matrix **P** required.

The method of alternately scaling the rows and columns of a matrix with a given structure of cross-product ratios, until specified row and column totals are attained, is known as the *iterative scaling procedure* (ISP) or iterative proportional fitting procedure (IPFP). A formal statement of the procedure for $r \times s$ tables is as follows, in which the notation for sums is to replace by zero any suffix over which summation takes place. We begin by scaling the rows of **Q** to meet the specified row totals. This stage of the iteration leads to

$$p_{ij}^{(1)} = c_i^{(1)}d_j^{(0)}q_{ij},$$

where $$c_i^{(1)} = p_{io}/q_{io} \quad \text{and} \quad d_j^{(0)} = 1.$$

The next stage is to scale the columns of $\{p_{ij}^{(1)}\}$ to meet the specified column totals. We obtain

$$p_{ij}^{(2)} = c_i^{(1)}d_j^{(1)}q_{ij},$$

where $$d_j^{(1)} = p_{oj}/p_{oj}^{(1)}.$$

In general, when the rows have been scaled n times and the columns $n-1$ times the result is

$$p_{ij}^{(2n-1)} = c_i^{(n)}d_j^{(n-1)}q_{ij},$$

where $$c_i^{(n)} = (p_{io}/p_{io}^{(2n-2)})c_i^{(n-1)}.$$

When rows and columns have each been scaled n times,

$$p_{ij}^{(2n)} = c_i^{(n)}d_j^{(n)}q_{ij},$$

where $$d_j^{(n)} = (p_{oj}/p_{oj}^{(2n-1)})\,d_j^{(n-1)}.$$

The sequences $\{c_i^{(n)}\}$, $\{d_j^{(n)}\}$, and $\{p_{ij}^{(n)}\}$ converge respectively to values c_i, d_j, and p_{ij} such that

$$p_{ij} = c_id_iq_{ij}.$$

An account of the iterative scaling procedure by Fienberg (1970b) relates the procedure to a variety of sources and a variety of statistical problems.

4 2 X 2 TABLES

4.1 Hypergeometric distribution

When we observe a binary response in each of two populations, or two binary responses in a single population, the results obtained by sampling can be presented in a 2 X 2 contingency table. Write

$$\lambda = \log(p_{11}p_{22}/p_{12}p_{21}).$$

For each experimental situation, inferences about λ are based on the distribution with probabilities

$$h(n_{11}; n_{io}, n_{oj}, \lambda) = \{\exp(n_{11}\lambda)/\Pi_{ij}\, n_{ij}!\}/f(\lambda),$$

where

$$f(\lambda) = \sum_{n_{11}=v}^{w} \exp(n_{11}\lambda)/\Pi_{ij}\, n_{ij}!$$

The limits of summation are

$$v = \max(0, n_{10} + n_{01} - n) \quad \text{and} \quad w = \min(n_{10}, n_{01}).$$

In this section, we consider the distribution when $\lambda = 0$. The probabilities are given in terms of the frequencies $\{n_{ij}\}$ and

$$f(0) = \sum_{n_{11}=v}^{w} (1/\Pi_{ij}\, n_{ij}!).$$

A simple expression for $f(0)$ can be derived by evaluating the coefficient of $x^{n_{01}}$ on both sides of the identity

$$(1+x)^{n_{10}}(1+x)^{n_{20}} \equiv (1+x)^n.$$

This shows that

$$\sum_{n_{11}=v}^{w} \binom{n_{10}}{n_{11}}\binom{n_{20}}{n_{21}} = \binom{n}{n_{01}},$$

whence we obtain

$$\sum_{n_{11}=v}^{w} (1/\Pi_{ij}\, n_{ij}!) = n!/n_{10}!\, n_{20}!\, n_{01}!\, n_{02}!$$

Consequently $h(n_{11}; n_{io}, n_{oj}, 0)$ can be written in either of the equivalent versions

$$\binom{n_{10}}{n_{11}}\binom{n_{20}}{n_{21}}\Big/\binom{n}{n_{01}} \quad \text{and} \quad \binom{n_{01}}{n_{11}}\binom{n_{02}}{n_{12}}\Big/\binom{n}{n_{10}}.$$

We say that the probabilities $\{h(n_{11}; n_{io}, n_{oj}, 0)\}$ form a *hypergeometric* distribution. The name arises because such quantities appear as coefficients in one of the series solutions of the hypergeometric equation.

The factorial moments of $h(n_{11}; n_{io}, n_{oj}, 0)$ are obtained by the following procedure. We take the distribution in the form

$$h(n_{11}; n_{io}, n_{oj}, 0) = n_{10}!\, n_{20}!\, n_{01}!\, n_{02}!/n!\, n_{11}!\, n_{12}!\, n_{21}!\, n_{22}!,$$

and hence obtain

$$n_{11}^{(r)} h(n_{11}; n_{io}, n_{oj}, 0) = \{n_{10}^{(r)} n_{01}^{(r)}/n^{(r)}\} h(n_{11}-r; n_{10}-r, n_{20}, n_{01}-r, n_{02}, 0),$$

where $x^{(r)}$ stands for $x(x-1)(x-2)\ldots(x-r+1)$. Therefore

$$E\{N_{11}^{(r)}\} = n_{10}^{(r)} n_{01}^{(r)}/n^{(r)}.$$

In this way – or by other methods – we find that the first three cumulants of N_{11} are given by the expressions below (Kendall and Stuart, vol. 1, ch. 5).

$$\kappa_1 = n_{10}n_{01}/n, \quad \kappa_2 = n_{10}n_{20}n_{01}n_{02}/n^2(n-1),$$

and

$$\kappa_3 = n_{10}n_{01}(n-n_{10})(n-n_{01})(n-2n_{10})(n-2n_{01})/n^3(n-1)(n-2).$$

Further properties of the hypergeometric distribution appear as special cases of results in the section which follows.

4.2 Extended hypergeometric distribution

We now consider the distribution with probabilities $\{h(n_{11}; n_{io}, n_{oj}, \lambda)\}$ when the value of λ is unrestricted. In view of the fact that N_{11} has a hypergeometric distribution when $\lambda = 0$, we say that N_{11} has an *extended hypergeometric* distribution.

Denote the cumulants of N_{11} by $\kappa_j(\lambda)$ and let the cumulant-generating function be

$$K(u;\lambda) = \log E\{\exp(uN_{11})\}.$$

This function is expressed in terms of $f(.)$ by

$$K(u;\lambda) = \log f(\lambda+u) - \log f(\lambda).$$

A consequence is that

$$\frac{\partial K}{\partial u} = \frac{\partial K}{\partial \lambda} + \frac{d\log f}{d\lambda}.$$

Equating the coefficients of powers of u, we obtain

$$\kappa_1(\lambda) = d\log f/d\lambda,$$

and

$$\kappa_j(\lambda) = d\kappa_{j-1}/d\lambda \quad (j = 2, 3, \ldots).$$

These results hold generally for the exponential family (Kendall and Stuart, vol. 2, eq. 23.107). In what follows, denote the mean and variance of N_{11} by $E\,(N_{11};\lambda)$ and var $(N_{11};\lambda)$ respectively. An expansion of each in powers of λ gives

$$E\,(N_{11};\lambda) = \kappa_1 + \lambda\kappa_2 + O(\lambda^2),$$

and

$$\operatorname{var}(N_{11};\lambda) = \kappa_2 + \lambda\kappa_3 + O(\lambda^2).$$

These expressions are due to Cox (1958). They are useful only for values of λ close to zero.

The moments and distribution function of N_{11} are the ratios of polynomials in e^{λ}. For moderate values of $\{n_{ij}\}$, the exact expressions are inconvenient to use without good computational facilities, and for large values of $\{n_{ij}\}$ they eventually become useless. Approximate expressions can be derived from the asymptotic properties of the distribution of N_{11}, which we now proceed to discuss. There are three forms of the asymptotic distribution depending on the way that n_{10}, n_{01} and $n \to \infty$ (Harkness, 1965). First, suppose that n_{10}, n_{01} and $n \to \infty$ such that $n_{10}/n \to \alpha$, where $0 < \alpha < 1$. Denote $\alpha e^{\lambda}/(1-\alpha+\alpha e^{\lambda})$ by β. Then

$$\lim h(n_{11}; n_{i0}, n_{0j}, \lambda) = \binom{n_{01}}{n_{11}} \beta^{n_{11}}(1-\beta)^{n_{21}}.$$

This generalizes the binomial limit of a hypergeometric distribution. Secondly, suppose that n_{10}, n_{01} and $n \to \infty$ such that $n_{10}n_{01}/n \to \xi$. Denote ξe^{λ} by μ. Then

$$\lim h(n_{11}; n_{i0}, n_{0j}, \lambda) = \mu^{n_{11}} e^{-\mu}/n_{11}!$$

This generalizes the Poisson limit of a hypergeometric distribution. The proofs of the binomial and Poisson limits are straightforward extensions of those for $\lambda = 0$. We now turn to the generalized normal limit.

There is a unique set of quantities $\{m_{ij}\}$ which satisfy the conditions

$$\Sigma_j\, m_{ij} = n_{i0}, \qquad \Sigma_i\, m_{ij} = n_{0j},$$

and

$$\log\,(m_{11}m_{22}/m_{12}m_{21}) = \lambda.$$

Suppose that $\{n_{i0}\}$ and $\{n_{0j}\}$ tend to infinity in such a way that $n_{10}/n \to \pi_{10}$ and $n_{01}/n \to \pi_{01}$, where

$$0 < \pi_{10} < 1 \quad \text{and} \quad 0 < \pi_{01} < 1.$$

Consider the variate

$$X = (N_{11} - m_{11})(m_{11}^{-1} + m_{12}^{-1} + m_{21}^{-1} + m_{22}^{-1})^{\frac{1}{2}}.$$

Then X converges in distribution to $\mathfrak{N}\,(0, 1)$ as $n \to \infty$. A proof is outlined by Cornfield (1956) and another is given in detail by Hannan and Harkness (1963).

The following proof is based on Cornfield's outline. Define π_{20} and π_{02} by

$$\pi_{10} + \pi_{20} = 1 \quad \text{and} \quad \pi_{01} + \pi_{02} = 1.$$

Introduce positive quantities $\{\pi_{ij}\}$ such that

$$\Sigma_j \, \pi_{ij} = \pi_{i0}, \quad \Sigma_i \, \pi_{ij} = \pi_{0j}, \quad \text{and} \quad \log(\pi_{11}\pi_{22}/\pi_{12}\pi_{21}) = \lambda.$$

Denote $h(n_{11}; n_{i0}, n_{0j}, \lambda)$ by $q(n_{11})$. This probability can be defined for all positive real values of $\{n_{ij}\}$ by using the factorial function. We eliminate $f(\lambda)$ by considering the ratio

$$\frac{q(n_{11})}{q(m_{11})} = \frac{m_{11}!\, m_{12}!\, m_{21}!\, m_{22}!}{n_{11}!\, n_{12}!\, n_{21}!\, n_{22}!} \, e^{\lambda(n_{11}-m_{11})}.$$

Each value of $c!$ is now replaced by Stirling's formula $(2\pi)^{\frac{1}{2}} c^{c+\frac{1}{2}} c^{-c}$. The contributions from $(2\pi)^{\frac{1}{2}}$ cancel out, as do those from e^{-c}. Consider next the contributions to $q(n_{11})/q(m_{11})$ from $c^{c+\frac{1}{2}}$. They are more conveniently treated as contributions to $\log q(n_{11}) - \log q(m_{11})$ from $(c + \frac{1}{2}) \log c$. Substitute $m_{11} + d$ for n_{11}, whence

$$(n_{11} + \tfrac{1}{2}) \log n_{11} = (n_{11} + \tfrac{1}{2}) \log m_{11} + (m_{11} + d + \tfrac{1}{2}) \log (1 + d/m_{11}).$$

We expand the second term on the right, assuming that d is $O(n^{\frac{1}{2}})$, and subsequently show that the probabilities from other values of d are negligible. Since $m_{ij} \sim n\pi_{ij}$,

$$(n_{11} + \tfrac{1}{2}) \log n_{11} = (n_{11} + \tfrac{1}{2}) \log m_{11} + d + d^2/2m_{11} + O(n^{-\frac{1}{2}}),$$

and hence $\qquad \log m_{11}! - \log n_{11}! \sim -d \log m_{11} - d - d^2/2m_{11}.$

The definition of d implies that

$$n_{12} = m_{12} - d, \quad n_{21} = m_{21} - d, \quad n_{22} = m_{22} + d.$$

Expansions of $(c + \frac{1}{2}) \log c$ similar to that when $c = n_{11}$ are therefore obtainable when $c = n_{12}, n_{21}$ and n_{22}. When the various contributions to $\log q\,(n_{11}) - \log q\,(m_{11})$ are collected, the coefficient of d is zero, and that of d^2 is

$$-\tfrac{1}{2}(m_{11}^{-1} + m_{12}^{-1} + m_{21}^{-1} + m_{22}^{-1}).$$

Consequently,

$$q(n_{11}) \sim q(m_{11}) \exp \{-\tfrac{1}{2} d^2 (m_{11}^{-1} + m_{12}^{-1} + m_{21}^{-1} + m_{22}^{-1})\}.$$

The sum of all such probabilities can be replaced by an integral, whence

$$q(m_{11}) \sim \{2\pi\,(m_{11}^{-1} + m_{12}^{-1} + m_{21}^{-1} + m_{22}^{-1})^{-1}\}^{-\frac{1}{2}}.$$

This shows that X is asymptotically $\mathfrak{N}(0, 1)$, and confirms that values of d larger than $O(n^{\frac{1}{2}})$ can be neglected.

An approximation to the distribution function of N_{11} is

$$p(N_{11}\leqslant n_{11})\simeq \Phi\,\{(n_{11}+\tfrac{1}{2}-m_{11})(m_{11}^{-1}+m_{12}^{-1}+m_{21}^{-1}+m_{22}^{-1})^{\frac{1}{2}}\},$$

where $\Phi(.)$ is the distribution function of $\mathfrak{N}\,(0, 1)$.

When $\lambda = 0$, the equations which define $\{m_{ij}\}$ have the solution

$$m_{ij} = n_{i0}n_{0j}/n,$$

in which case

$$m_{11}^{-1}+m_{12}^{-1}+m_{21}^{-1}+m_{22}^{-1} = n^3/n_{10}n_{20}n_{01}n_{02}.$$

We now have

$$X = (N_{11}-n_{10}n_{01}/n)/(n_{10}n_{20}n_{01}n_{02}/n^3)^{\frac{1}{2}},$$

which differs only slightly from the standardized variate $(N_{11} - \kappa_1)/\kappa_2^{\frac{1}{2}}$.

Another normal approximation to the distribution of N_{11} when $\lambda = 0$ is given by Nicholson (1956). The asymptotic distribution of N_{11} derived in this section will be generalized in what follows, first when we discuss the distribution theory of $r \times s$ tables in section 5.1 and then again with reference to three-dimensional tables in section 7.3. Haberman (1970, ch. 1) gives a rigorous statement and proof of a theorem assuring the normal convergence of all such conditional Poisson distributions.

4.3 Sample cross-product ratio

The methods of inference which are based on the extended hypergeometric distribution involve not only N_{11} but also the sample equivalent of the odds or cross-product ratio, namely $N_{11}N_{22}/N_{12}N_{21}$. Write

$$L = \log(N_{11}N_{22}/N_{12}N_{21}).$$

L is a function of N_{11} because of the conditions

$$N_{10} = n_{10}, \quad N_{01} = n_{01}, \quad \text{and} \quad N = n,$$

where $N = \Sigma_{ij} N_{ij}$. Consider the variate

$$V = (L-\lambda)(m_{11}^{-1}+m_{12}^{-1}+m_{21}^{-1}+m_{22}^{-1})^{-\frac{1}{2}},$$

where the quantities $\{m_{ij}\}$ are uniquely defined by

$$m_{11}+m_{12} = n_{10}, \quad m_{11}+m_{21} = n_{01}, \quad \Sigma_{ij}\, m_{ij} = n,$$

and

$$\log(m_{11}m_{22}/m_{12}m_{21}) = \lambda.$$

If the conditions under which X converges to $\mathfrak{N}\,(0, 1)$ are satisfied, then V also converges to $\mathfrak{N}\,(0, 1)$. This result can be obtained directly from the asymptotic normality of X, but a more illuminating derivation is as follows.

Suppose that the variates $\{N_{ij}\}$ are mutually independent and such that N_{ij} has a Poisson distribution with parameter $\mu\pi_{ij}$, where

$$0<\pi_{ij}<1 \quad \text{and} \quad \Sigma_{ij}\,\pi_{ij} = 1.$$

As before, write

$$\pi_{10} = \pi_{11} + \pi_{12}, \quad \pi_{01} = \pi_{11} + \pi_{21},$$

and

$$\lambda = \log(\pi_{11}\pi_{22}/\pi_{12}\pi_{21}).$$

Introduce fixed numbers t_{10}, t_{01} and t. Assign to n_{10}, n_{01} and n the values

$$n_{10} = [\mu\pi_{10} + t_{10}(\mu\pi_{10})^{\frac{1}{2}}], \quad n_{01} = [\mu\pi_{01} + t_{01}(\mu\pi_{01})^{\frac{1}{2}}],$$

and

$$n = [\mu + t\mu^{\frac{1}{2}}].$$

In these equations only, $[x]$ stands for x when x is an integer, and otherwise the integer next below x. Consider the distribution of V conditional on $N_{10} = n_{10}, N_{01} = n_{01}$, and $N = n$ in the limit as $\mu \to \infty$. Then the conditional distribution of V converges to $\mathfrak{N}(0, 1)$.

An outline of the proof is as follows. The transformation from $\{N_{ij}\}$ to N, N_{10}, N_{01} and L has a unique inverse because of the result that the entries in a contingency table are uniquely determined by the marginal totals and cross-product ratios. Thus the joint distribution of N, N_{10}, N_{01} and L is non-singular. The distribution of $(N_{ij} - \mu\pi_{ij})/(\mu\pi_{ij})^{\frac{1}{2}}$ converges to $\mathfrak{N}(0, 1)$ as $\mu \to \infty$. We deduce in succession that the same is true for

$$(\mu\pi_{ij})^{\frac{1}{2}}\{\log N_{ij} - \log(\mu\pi_{ij})\},$$

and for

$$U = \mu^{\frac{1}{2}}(L - \lambda)(\pi_{11}^{-1} + \pi_{12}^{-1} + \pi_{21}^{-1} + \pi_{22}^{-1})^{-\frac{1}{2}}.$$

However, the asymptotic distribution of U conditional on $N_{10} = n_{10}, N_{01} = n_{01}$ and $N = n$ is the same as the unconditional distribution because U is asymptotically independent of N_{10}, N_{01} and N. In view of the asymptotic normality of all the variates concerned, asymptotic independence is established by showing that all the covariances between L and the other variates are asymptotically zero. Consider, for example, the asymptotic covariance of L and N_{10}. This is obtained by expanding the logarithms about their mean values, and is given by

$$\operatorname{cov}(\log N_{11} - \log N_{12}, N_{11} + N_{12})$$
$$\sim \operatorname{cov}\{(N_{11} - \mu\pi_{11})/\mu\pi_{11} - (N_{12} - \mu\pi_{12})/\mu\pi_{12}, N_{11} + N_{12}\} = 0.$$

All other covariances are similarly zero. The proof is completed by observing that U and V are asymptotically equal.

4.4 Estimation of λ

The parameter λ is defined in terms of the probabilities of a 2×2 contingency table by

$$\lambda = \log(p_{11}p_{22}/p_{12}p_{21}).$$

When there are two populations in which a binary response is observed, λ

measures the logit difference between the probabilities of success; and when there is one population in which two binary responses are observed, λ measures the degree of association.

We turn now to the methods by which inferences about λ are made. The other parameters which arise in connection with the analysis of 2 X 2 tables are

$$\kappa = \log p_{22}, \quad \theta = \log (p_{12}/p_{22}), \quad \text{and} \quad \phi = \log (p_{21}/p_{22}).$$

In each of the situations which give rise to the table, there are two likelihood functions involving λ. One is the unconditional likelihood function obtained from the distribution of $\{N_{ij}\}$, and given by

$$\exp (n_{10}\theta + n_{11}\lambda)(n_{01}!\, n_{02}!\, p_{21}^{n_{01}} p_{22}^{n_{02}}/\Pi_{ij}\, n_{ij}!)$$

or

$$\exp (n\kappa + n_{10}\theta + n_{01}\phi + n_{11}\lambda)(n!/\Pi_{ij}\, n_{ij}!),$$

according as one or both axes of the table refer to responses. The other is the conditional likelihood function obtained from $h(n_{11}; n_{i0}, n_{0j}, \lambda)$. There are the following reasons, which are respectively mathematical and philosophical, for preferring to use the second likelihood function to make inferences about λ. First, the properties of multiparameter exponential families of distributions, to which the distributions of $\{N_{ij}\}$ belong, imply that uniformly most powerful unbiased tests of hypotheses about λ are derived from the conditional distribution (Lehmann, 1959, ch. 4). Secondly, the marginal totals of the table cannot by themselves supply any information about the proportionality of the frequencies in the body of the table, but only ancillary information about the precision realized. We should therefore exclude from consideration marginal totals which have not been observed (Fisher, 1935; 1959, sect. IV.4).

Consider first the point estimation of λ from an observation on N_{11}. Denote the conditional maximum-likelihood estimator by $\tilde{\lambda}$. The first two derivatives of the log likelihood function are

$$\frac{\partial \log h}{\partial \lambda} = n_{11} - \frac{1}{f}\frac{\partial f}{\partial \lambda}$$

and

$$\frac{\partial^2 \log h}{\partial \lambda^2} = -\frac{1}{f}\frac{\partial^2 f}{\partial \lambda^2} + \left(\frac{1}{f}\frac{\partial f}{\partial \lambda}\right)^2.$$

Thus $\tilde{\lambda}$ satisfies the equation

$$N_{11} - E(N_{11}; \tilde{\lambda}) = 0,$$

and

$$\operatorname{var}(\tilde{\lambda}) \sim 1/\operatorname{var}(N_{11}; \lambda).$$

The left-hand side of the equation for $\tilde{\lambda}$ is a polynomial of degree $w - v$ in $\tilde{\psi} = \exp(\tilde{\lambda})$. There is a unique solution of the equation because the derivative of $E(N_{11};\lambda)$ is var $(N_{11};\lambda)$, which is positive for all λ. The solution can be found by numerical methods, for example the Newton–Raphson procedure, where a sequence of iterates is constructed in the following way. Denote $n_{11} - E(N_{11};\log\psi)$ by $g(\psi)$. Let ψ_k be the kth approximation to $\tilde{\psi}$. Then

$$\psi_{k+1} = \psi_k - g(\psi_k)/g'(\psi_k).$$

The sequence thus obtained is *second-order*, in the sense that

$$|\psi_{k+1} - \tilde{\psi}| = c|\psi_k - \tilde{\psi}|^2,$$

where c is a positive constant.

Alternatively, the solution can be found from approximate formulae. When λ is small, we can replace $E(N_{11};\lambda)$ by $\kappa_1 + \lambda\kappa_2$, and hence obtain (Birch, 1964) the approximate estimator

$$\lambda^* = (N_{11} - \kappa_1)/\kappa_2.$$

An expansion of $E(\lambda^*)$ in powers of λ is given by Goodman (1969). He derives an estimator λ^{**} which improves on λ^* when λ differs somewhat from zero.

When the likelihood function is either of the unconditional versions given at the beginning of this section, the maximum-likelihood estimator of λ is

$$\hat{\lambda} = \log(N_{11}N_{22}/N_{12}N_{21}).$$

At least one of the frequencies can take the value zero, and therefore $E(\hat{\lambda})$ does not exist. The difficulty can be removed by using instead the estimator

$$\lambda^+ = \log\{(N_{11}+\tfrac{1}{2})(N_{22}+\tfrac{1}{2})/(N_{12}+\tfrac{1}{2})(N_{21}+\tfrac{1}{2})\}.$$

Consider the use of λ^+ as an estimator of λ in the conditional distribution. The arguments used in the previous section imply that the asymptotic value of $E(\lambda^+)$ can be derived by supposing that the $\{N_{ij}\}$ are mutually independent Poisson variates with parameters $\{\mu\pi_{ij}\}$ respectively. A reference to section 1.2 shows that

$$E\{\log(N_{ij}+\tfrac{1}{2})\} = \log(\mu\pi_{ij}) + O(\mu^{-2}),$$

and therefore

$$E(\lambda^+) = \lambda + O(\mu^{-2}).$$

Numerical details about the bias of λ^+ in the case of samples from two binomial distributions are given by Gart and Zweifel (1967). Their results support the estimation of var (λ^+) by

$$(N_{11}+\tfrac{1}{2})^{-1} + (N_{12}+\tfrac{1}{2})^{-1} + (N_{21}+\tfrac{1}{2})^{-1} + (N_{22}+\tfrac{1}{2})^{-1}.$$

Example 4.1 The data appear in Table 3.1. Here $\lambda^* = 1{\cdot}5844$. $\tilde{\psi}$ satisfies the equation

$$\sum_{x=0}^{9} (x-4)\binom{16}{x}\binom{79}{9-x}\tilde{\psi}^x \Big/ \binom{95}{9} = 0.$$

Successive approximations give $\tilde{\psi} = 4{\cdot}8147$, whence $\tilde{\lambda} = 1{\cdot}5717$.

Consider next the construction of confidence intervals for λ. Exact confidence limits can be obtained by the standard method for one-parameter families of distributions. For a specified probability ϵ, define λ_2 and λ_1 respectively by

$$\sum_{x=v}^{n_{11}} h(x; n_{i0}, n_{0j}, \lambda_2) = \tfrac{1}{2}\epsilon, \qquad \sum_{x=n_{11}}^{w} h(x; n_{i0}, n_{0j}, \lambda_1) = \tfrac{1}{2}\epsilon.$$

Then (λ_1, λ_2) is a confidence interval for λ with a confidence coefficient at least $1-\epsilon$. The relevant calculations are described by Cox (1970a, sect. 4.3). They have been programmed by Thomas (1971), Mantel and Hankey (1971), and Evans (1973). Noncentral intervals can be obtained in the same way. The best choice of tail probabilities depends on the criterion for optimality.

Approximate confidence intervals in large samples can be derived from the asymptotic properties of N_{11}. Three asymptotic distributions are defined in section 4.3, and the choice of distribution is guided by the magnitudes of the marginal frequencies. Suppose, for example, that the asymptotic distribution is taken as normal. Denote by u_ϵ the upper 100ϵ percent point of $\mathfrak{N}(0, 1)$. The equation

$$(n_{11} + \tfrac{1}{2} - m_{11})(m_{11}^{-1} + m_{12}^{-1} + m_{21}^{-1} + m_{22}^{-1})^{\frac{1}{2}} = -u_{\frac{1}{2}\epsilon}$$

leads to a quartic in m_{11}, and the smallest real root determines a lower confidence limit for λ. An upper confidence limit is similarly obtained, and the two limits define a confidence interval with confidence coefficient approximately $1-\epsilon$. This method was presented by Cornfield (1956) and independently by Fisher (1962).

Another approach is to use the asymptotic properties of $\hat{\lambda}$ or λ^+. Under the conditions given in the previous section, where $\hat{\lambda}$ was denoted by L, the variate

$$V = (\hat{\lambda} - \lambda)(m_{11}^{-1} + m_{12}^{-1} + m_{21}^{-1} + m_{22}^{-1})^{-\frac{1}{2}}$$

converges to $\mathfrak{N}(0, 1)$. The same result holds when m_{ij} is replaced by N_{ij} because $|N_{ij} - m_{ij}|$ is $O(n^{-\frac{1}{2}})$ with a probability which tends to 1 as $n \to \infty$. An additional change from N_{ij} to $N_{ij} + \frac{1}{2}$, for the reasons given above, shows that the limits

$$\log\{(n_{11} + \tfrac{1}{2})(n_{22} + \tfrac{1}{2})/(n_{12} + \tfrac{1}{2})(n_{21} + \tfrac{1}{2})\}$$
$$\pm u_{\frac{1}{2}\epsilon}\{(n_{11} + \tfrac{1}{2})^{-1} + (n_{12} + \tfrac{1}{2})^{-1} + (n_{21} + \tfrac{1}{2})^{-1} + (n_{22} + \tfrac{1}{2})^{-1}\}^{\frac{1}{2}}$$

form a confidence interval for λ with confidence coefficient approximately $1 - \epsilon$. This procedure is known as the "logit" method.

A third method is based on a binomial approximation to the hypergeometric distribution (Gart, 1962). Various methods of calculating approximate confidence limits are studied numerically by Gart and Thomas (1972) for tables where the marginal totals are 10, 15 or 20. They find that the best of the approximate methods are those due to Cornfield and Gart, but the limits obtained from the logit method are much too narrow.

Example 4.2 The data appear in Table 3.1. For a confidence coefficient of 0·95, the program constructed by Thomas (1971) gives $0{\cdot}8341 \leqslant \psi \leqslant 26{\cdot}16$.

4.5 Hypotheses about λ

A test of the hypothesis $\lambda \leqslant 0$ against the alternative $\lambda > 0$ at significance level ϵ is obtained as follows. Define c as the smallest value such that

$$\sum_{x=c}^{w} h(x; n_{i0}, n_{0j}, 0) \leqslant \epsilon.$$

The critical region consists of the values $c, c+1, \ldots, w$ of n_{11}. Values of c for significance levels of 0·05, 0·025, 0·01, and 0·005, together with the exact size of the test, have been computed by Finney *et al.* (1963) and are reproduced by Pearson and Hartley (1966, Table 38). A test of the hypothesis $\lambda = 0$ against the alternative $\lambda \neq 0$ is similarly made by using a critical region which consists of both large and small values of n_{11}. These procedures were first proposed by Fisher (1934, sect. 21.02) and Irwin (1935) to test the hypothesis of independent responses.

Returning to the one-sided test, we may use a randomizing device to include $n_{11} = c - 1$ in the critical region with a probability which makes the size of the test exactly ϵ. The procedure thus obtained is a uniformly most powerful unbiased test of $\lambda \leqslant 0$ against $\lambda > 0$. A test of $\lambda = 0$ against $\lambda \neq 0$ with the same optimum property is similarly constructed. These results are due to Tocher (1950) and they also follow from the theory of tests for multiparameter exponential families given by Lehmann (1959, ch. 4). As already noted in section 2.2, randomization is seldom used in practice.

Example 4.3 The data appear in Table 3.1. An exact test of the hypothesis $\psi \leqslant 1$ against the alternative $\psi > 1$ calls for the sum of the probabilities corresponding to the tables below.

4	5	5	4	6	3	7	2	8	1	9	0
12	74	11	75	10	76	9	77	8	78	7	79

The total probability is 0·04106, so that $\psi \leqslant 1$ is rejected by a one-sided test with a significance level of 0·05. We now conclude the analysis of Table 3.1. Avadex increases the logit of a pulmonary tumour by an estimated 1·6, but the confidence interval is wide and the increase is only just confirmed by a one-sided test.

When the marginal totals are large, we proceed as follows. Suppose that the problem is to test the hypothesis

$$H : \log (p_{11}p_{22}/p_{12}p_{21}) = \lambda_0$$

against the alternative

$$K : \log (p_{11}p_{22}/p_{12}p_{21}) \neq \lambda_0,$$

where the value λ_0 is specified. There is a unique set of quantities $\{m_{ij}(\lambda_0)\}$ which satisfy the conditions

$$\Sigma_j\, m_{ij}(\lambda_0) = n_{io}, \quad \Sigma_i\, m_{ij}(\lambda_0) = n_{oj},$$

and $$\log \{m_{11}(\lambda_0)\, m_{22}(\lambda_0)/m_{12}(\lambda_0)\, m_{21}(\lambda_0)\} = \lambda_0.$$

If the likelihood function is either of the unconditional versions given at the beginning of the previous section, then $m_{ij}(\lambda_0)$ is the maximum-likelihood estimate of np_{ij} under H. For example, consider the maximization of $\Sigma_{ij}\, n_{ij} \log p_{ij}$ subject to $\Sigma_{ij}\, p_{ij} = 1$ and H. The maximum is obtained by introducing Lagrangian multipliers β and δ, and differentiating

$$\Sigma_{ij}\, n_{ij} \log p_{ij} - \beta\, \Sigma_{ij}\, p_{ij} - \delta\, (\log p_{11} - \log p_{12} - \log p_{21} + \log p_{22})$$

with respect to $\{p_{ij}\}$. We find that $\beta = n$, whence

$$n\hat{p}_{io} = n_{io} \quad \text{and} \quad n\hat{p}_{oj} = n_{oj}.$$

Consequently $n\hat{p}_{ij} = m_{ij}(\lambda_0)$ as stated. We say that np_{ij} is an *expected frequency*. By tradition, the same term is attached to the estimate $m_{ij}(\lambda_0)$, as already noted in section 1.3.

A continuation of this line of approach leads to two large-sample tests for either of the unconditional likelihood functions. The first procedure is based on Karl Pearson's statistic

$$\Sigma\, (\text{Observed} - \text{Expected})^2/\text{Expected},$$

which gives the test statistic

$$X^2 = \Sigma_{ij}\, \{N_{ij} - M_{ij}(\lambda_0)\}^2/M_{ij}(\lambda_0).$$

An alternative is minus twice the logarithm of the likelihood ratio, which gives the test statistic

$$Y^2 = 2\, \Sigma_{ij}\, N_{ij} \log \{N_{ij}/M_{ij}(\lambda_0)\}.$$

Both X^2 and Y^2 converge to χ_1^2 when H is true.

We now show that the same statistics can be derived from the conditional likelihood function. An asymptotic version of the optimum test procedure is to refer N_{11} to a normal distribution with mean $m_{11}(\lambda_0)$ and variance

$$\{m_{11}^{-1}(\lambda_0) + m_{12}^{-1}(\lambda_0) + m_{21}^{-1}(\lambda_0) + m_{22}^{-1}(\lambda_0)\}^{-1}$$

Equivalently, we can refer the statistic

$$X^2 = \{N_{11} - m_{11}(\lambda_0)\}^2 \{m_{11}^{-1}(\lambda_0) + m_{12}^{-1}(\lambda_0) + m_{21}^{-1}(\lambda_0) + m_{22}^{-1}(\lambda_0)\}$$

to a table of χ_1^2. The system of equalities

$$N_{11} - m_{11}(\lambda_0) = -\{N_{12} - m_{12}(\lambda_0)\} = -\{N_{21} - m_{21}(\lambda_0)\} = N_{22} - m_{22}(\lambda_0)$$

implies that

$$X^2 = \Sigma_{ij} \{N_{ij} - m_{ij}(\lambda_0)\}^2 / m_{ij}(\lambda_0).$$

Alternatively, we can begin with the exact likelihood ratio

$$T = h(N_{11}; n_{i0}, n_{0j}, \lambda_0) / h(N_{11}; n_{i0}, n_{0j}, \tilde{\lambda}).$$

This is an inconvenient statistic in small samples because of the need to calculate $\tilde{\lambda}$. An asymptotic version is obtained as follows. The definition of $\tilde{\lambda}$ implies that the denominator of T is

$$h\{E(N_{11}; \tilde{\lambda}); n_{i0}, n_{0j}, \tilde{\lambda}\}.$$

When H is true, $\tilde{\lambda}$ converges in probability to λ_0, and a theorem of Cramér (1946, sect. 20.6) then ensures that T has the same asymptotic distribution as

$$U = h(N_{11}; n_{i0}, n_{0j}, \lambda_0) h\{m_{11}(\lambda_0); n_{i0}, n_{0j}, \lambda_0\}.$$

Consider the statistic $-2 \log U$. Under the conditions stated in section 4.2, the arguments given there imply that $-2 \log U \sim X^2$ and hence converges to χ_1^2 when H is true. Furthermore,

$$-2 \log U = -2\, \Sigma_{ij} \{m_{ij}(\lambda_0) + \tfrac{1}{2}\} \log m_{ij}(\lambda_0) + 2\, \Sigma_{ij} (N_{ij} + \tfrac{1}{2}) \log N_{ij} - $$
$$- 2\, \Sigma_{ij} \{N_{ij} - m_{ij}(\lambda_0)\} \log m_{ij}(\lambda_0).$$

Omission of the $\frac{1}{2}$ leads to the deletion of terms which are asymptotically negligible compared with those which remain, and the resulting test statistic is

$$Y^2 = 2\, \Sigma_{ij} N_{ij} \log \{N_{ij} / m_{ij}(\lambda_0)\}.$$

We conclude that appropriate large-sample statistics for testing H against K are X^2 and Y^2. They are asymptotically equivalent and the numerical difference will usually be small.

Example 4.4 The table opposite was given in section 3.3, and suggests a positive association between enlarged tonsils and carriers of *Streptococcus pyogenes*.

Tonsils	Carriers	Non-carriers	Total
Enlarged	53	829	882
Not enlarged	19	497	516
Total	72	1326	1398

We shall test the hypothesis that $\lambda = 0$ against the alternative $\lambda \neq 0$. A large-sample criterion is appropriate and the values of those put forward are as follows, where the probabilities are obtained from a table of χ_1^2.

$$X^2 = 3{\cdot}608\ (p = 0{\cdot}058); \quad Y^2 = 3{\cdot}778\ (p = 0{\cdot}051).$$

Each value confirms a slight association. When X^2 is corrected for continuity, the value becomes 3·148 ($p = 0{\cdot}076$). No correction is available for likelihood-ratio statistics.

4.6 Bayesian analysis

Suppose that the units of a single population are classified by two binary responses and allocated to the four cells (i, j) of a 2×2 table. We have assumed that the corresponding frequencies $\{N_{ij}\}$ in a sample of size n have a multinomial distribution with index n and probabilities $\{p_{ij}\}$. According to standard theory, the parameters $\{p_{ij}\}$ are fixed but unknown. In the Bayesian approach to inference they are the observed values of variates $\{P_{ij}\}$ to which a prior distribution is assigned. This may be a probability distribution in the ordinary sense, or an improper distribution with a non-negative density and infinite total integral. The likelihood of the observed frequencies $\{n_{ij}\}$ is combined with the prior distribution by Bayes's theorem to yield the distribution of $\{P_{ij}\}$ conditional on $\{N_{ij} = n_{ij}\}$, known as the "posterior distribution". This conditional distribution is the basis of Bayesian inferences about $\{p_{ij}\}$.

A convenient prior distribution for non-negative real numbers which sum to unity has the Dirichlet form with parameters $\{\nu'_{ij}\}$. Thus the probability density function of $\{P_{ij}\}$ is given by

$$f(p_{11}, p_{12}, p_{21}, p_{22}) = \Gamma(\Sigma_{ij}\, \nu'_{ij})\, \Pi_{ij}\, p_{ij}{}^{\nu'_{ij}-1}/\Gamma(\nu'_{ij}).$$

The likelihood of the frequencies $\{n_{ij}\}$ is

$$p(N_{ij} = n_{ij} \quad \text{for all } i, j) = n!\, \Pi_{ij}\, p_{ij}{}^{n_{ij}}/n_{ij}!$$

Consequently the posterior density function of $\{P_{ij}\}$ is

$$g(p_{11}, p_{12}, p_{21}, p_{22}) = \Gamma(\Sigma_{ij}\, \nu_{ij})\, \Pi_{ij}\, p_{ij}{}^{\nu_{ij}-1}/\Gamma(\nu_{ij}),$$

where $$\nu_{ij} = n_{ij} + \nu'_{ij} \quad \text{for all } i, j.$$

We have noted that a measure of association between responses should be a function of the cross-product ratio. Write

$$\Psi = P_{11}P_{22}/P_{12}P_{21}.$$

The posterior distribution of Ψ is somewhat intractable, but a simple expression for $p(\Psi < 1)$ is obtained as follows (Altham, 1969). Make a transformation from $\{P_{ij}\}$ to X, Y, T, and U, defined by

$$P_{11} = TX, \quad P_{12} = T(1-X), \quad P_{21} = UY, \quad P_{22} = U(1-Y).$$

The Jacobian of the transformation is

$$\partial(p_{11}, p_{12}, p_{21}, p_{22})/\partial(x, y, t, u) = tu.$$

Hence the density function of X, Y, T, and U is

$$\{B(\nu_{11}, \nu_{12})\}^{-1} x^{\nu_{11}-1}(1-x)^{\nu_{12}-1}\{B(\nu_{21}, \nu_{22})\}^{-1} y^{\nu_{21}-1}(1-y)^{\nu_{22}-1} \times$$
$$\times \{B(\nu_{10}, \nu_{20})\}^{-1} t^{\nu_{10}-1} u^{\nu_{20}-1},$$

where the notation for sums is the same as before, namely

$$\nu_{11} + \nu_{12} = \nu_{10} \quad \text{and} \quad \nu_{21} + \nu_{22} = \nu_{20}.$$

We deduce in particular that X and Y are independent. These variates are given by

$$X = P_{11}/P_{10} \quad \text{and} \quad Y = P_{21}/P_{20}.$$

Therefore the inequalities $\Psi < 1$ and $X < Y$ are equivalent, so that their probabilities are equal. Thus

$$p(\Psi < 1 \mid N_{ij} = n_{ij}, \quad \text{all } i, j)$$
$$= \int_0^1 \{B(\nu_{21}, \nu_{22})\}^{-1} y^{\nu_{21}-1}(1-y)^{\nu_{22}-1}$$
$$\times \left[\int_0^y \{B(\nu_{11}, \nu_{12})\}^{-1} x^{\nu_{11}-1}(1-x)^{\nu_{12}-1} dx\right] dy.$$

The integral in square brackets can be expressed as a sum of binomial probabilities in the form

$$\int_0^y \{B(\nu_{11}, \nu_{12})\}^{-1} x^{\nu_{11}-1}(1-x)^{\nu_{12}-1} = \sum_{r=\nu_{11}}^{\nu_{10}-1} \binom{\nu_{10}-1}{r} y^r(1-y)^{\nu_{10}-1-r}$$

One method of proof is to note that the integrand is the density function of the ν_{11}th order statistic in a sample of size $\nu_{10} - 1$ from the uniform distribution on (0, 1). Alternatively, integrate by parts. The probability required is now a linear combination of beta functions, whence

$$p(\Psi < 1 \mid n_{ij}, \quad \text{all } i, j)$$

$$= \sum_{r=\nu_{11}}^{\nu_{10}-1} \{B(\nu_{21}, \nu_{22})\}^{-1} \binom{\nu_{10}-1}{r} B(\nu_{21}+r, \nu-1-\nu_{21}-r),$$

in which ν denotes $\Sigma_{ij}\, \nu_{ij}$. Another expression is

$$p(\Psi < 1 \mid n_{ij}, \quad \text{all } i, j) = \sum_{s=\nu_{11}}^{\min(\nu_{10}-1,\, \nu_{01}-1)} h(s; \nu_{i0}-1, \nu_{0j}-1, 0).$$

We prove the equivalence of the two versions by considering a population containing $\nu_{10}-1$ successes and $\nu_{20}-1$ failures from which successive units are taken singly. The first version is the probability that at least ν_{11} successes occur before the ν_{21}th failure; and the second version is the probability that at least ν_{11} successes occur among the first $\nu_{01}-1$ units taken. Since these probabilities refer to the same event, they are equal.

Consider a table with the observed frequencies given below.

ν_{11}	$\nu_{12}-1$
$\nu_{21}-1$	ν_{22}

A test of the hypothesis $\lambda \leqslant 0$ against the alternative $\lambda > 0$, made by the method described in the previous section, would lead us to calculate the final expression for the posterior probability of the event $\Psi < 1$. Since we would reject $\lambda \leqslant 0$ if the probability were sufficiently small, the two approaches are consistent. The test of section 4.5 applied to a table with frequencies $\{n_{ij}\}$ would lead to the probability for $\Psi < 1$ if we take

$$\nu'_{11} = \nu'_{22} = 0 \quad \text{and} \quad \nu'_{12} = \nu'_{21} = 1.$$

This choice of parameters in the prior distribution corresponds to a strong prior belief that $\Psi < 1$.

Although the posterior distribution of Ψ is difficult to handle for finite values of the parameters, a normal approximation is available in the limit as $\nu_{ij} \to \infty$ for all i and j. Define $\Lambda = \log \Psi$,

$$\mu = \log(\nu_{11}\nu_{22}/\nu_{12}\nu_{21}) \quad \text{and} \quad \sigma^2 = \Sigma_{ij}\, \nu_{ij}^{-1}.$$

Then $(\Lambda - \mu)/\sigma$ converges in distribution to $\mathfrak{N}(0, 1)$ (Lindley, 1964). A modified version of this result is due to Bloch and Watson (1967), who show that the approximation for large $\{\nu_{ij}\}$ is improved if we take

$$\mu = \log(\tau_{11}\tau_{22}/\tau_{12}\tau_{21}) \quad \text{and} \quad \sigma^2 = \Sigma_{ij}\, \tau_{ij}^{-1},$$

where $$\tau_{ij} = \nu_{ij} - \tfrac{1}{2} + 1/24(\nu_{ij}-1).$$

5 $r \times s$ TABLES

5.1 Distribution theory

Each of the experimental situations which give rise to an $r \times s$ contingency table can be described by a statistical model based on rs mutually independent variates $\{N_{ij}\}$ such that N_{ij} has a Poisson distribution with parameter μ_{ij}, where $i = 1, 2, \ldots, r$ and $j = 1, 2, \ldots, s$. Write

$$\lambda_{ab} = \log(\mu_{ab}\mu_{rs}/\mu_{as}\mu_{rb}) \quad (a = 1, 2, \ldots, r-1;\ b = 1, 2, \ldots, s-1).$$

The arguments given in sections 3.2 and 3.3 can be summarized as follows. When a classification with r categories is made in s populations, the distribution of the response R is the same for each level of the factor $\mathfrak{F}$ if and only if $\lambda_{ab} = 0$ for all a and b. Again, when a population is classified by each of two methods which contain r and s categories respectively, the responses R and S are independent if and only if $\lambda_{ab} = 0$ for all a and b. Inferences about $\{\lambda_{ab}\}$ are based on the distribution of $\{N_{ab}\}$ conditional on

$$N = n, \quad N_{a0} = n_{a0}, \quad \text{and} \quad N_{0b} = n_{0b} \quad \text{for all } a \text{ and } b.$$

The distribution is denoted by $h\{n_{ab}; n_{i0}, n_{0j}, \lambda_{ab}\}$ and defined by

$$\{\exp(\Sigma_{ab}\, n_{ab}\lambda_{ab})/\Pi_{ij}\, n_{ij}!\}/\{\Sigma_{n_{ab}} \exp(\Sigma_{ab}\, n_{ab}\lambda_{ab})/\Pi_{ij}\, n_{ij}!\}.$$

We proceed to derive the main properties of $h\{.\}$. Most of the results are given only in outline, because they are obtained by direct generalization of the methods for 2×2 tables.

Consider first the case when $\lambda_{ab} = 0$ for all a and b. The distribution then takes the form

$$h\{n_{ab}; n_{i0}, n_{0j}, 0\} = (1/\Pi_{ij}\, n_{ij}!)/\{\Sigma_{n_{ab}} (1/\Pi_{ij}\, n_{ij}!)\}.$$

We evaluate the sum in the denominator by comparing the coefficient of $x_1^{n_{01}} x_2^{n_{02}} \ldots x_s^{n_{0s}}$ on both sides of the identity

$$(x_1 + x_2 + \ldots + x_s)^{n_{10}}(x_1 + x_2 + \ldots + x_s)^{n_{20}} \ldots (x_1 + x_2 + \ldots + x_s)^{n_{r0}} \equiv (x_1 + x_2 + \ldots + x_s)^n.$$

This shows that

$$\Sigma_{n_{ab}}\, \Pi_{ij}\, (n_{i0}!/n_{ij}!) = n!/\Pi_j\, n_{0j}!,$$

or equivalently

$$\Sigma_{n_{ab}}\, (1/\Pi_{ij}\, n_{ij}!) = n!/\Pi_{ij}\, n_{i0}!\, n_{0j}!$$

The distribution when $\lambda_{ab} = 0$ for all a and b is therefore

$$h\{n_{ab}; n_{i0}, n_{0j}, 0\} = \Pi_{ij}\,(n_{i0}!\,n_{0j}!/n_{ij}!)/n!$$

We can similarly generalize the method of deriving factorial moments given in section 4.1, and hence obtain the following results, which are due to Stevens (1938) and Haldane (1940).

$$E\,(N_{ij}) = n_{i0}n_{0j}/n,$$

and

$$\text{cov}\,(N_{ij}, N_{i'j'}) = n_{i0}(\delta_{ii'}\,n - n_{i'0})n_{0j}(\delta_{jj'}\,n - n_{0j'})/n^2(n-1).$$

The quantity δ_{ij} used here is such that

$$\delta_{ij} = \begin{cases} 1 & \text{if } i = j \\ 0 & \text{if } i \neq j. \end{cases}$$

Consider now the general case when the parameters $\{\lambda_{ab}\}$ are not all zero. Under the conditions which follow, the frequencies $\{N_{ab}\}$ have an asymptotic normal multivariate distribution. Let $\{\pi_{i0}\}$ and $\{\pi_{0j}\}$ be positive quantities such that

$$\Sigma_i\,\pi_{i0} = \Sigma_j\,\pi_{0j} = 1.$$

The parameters $\{\lambda_{ab}\}$ form a matrix of order $(r-1) \times (s-1)$. Write $\boldsymbol{\lambda}$ for the vector obtained when the columns of this matrix are adjoined to form a single column. Thus:

$$\boldsymbol{\lambda}^T = (\lambda_{11}, \lambda_{21}, \ldots, \lambda_{r-1,1}, \lambda_{12}, \lambda_{22}, \ldots, \lambda_{r-1,\,s-1}).$$

For a specified value of $\boldsymbol{\lambda}$, define $\{m_{ij}(\boldsymbol{\lambda})\}$ to satisfy

$$\Sigma_j\,m_{ij}(\boldsymbol{\lambda}) = n_{i0}, \quad \Sigma_i\,m_{ij}(\boldsymbol{\lambda}) = n_{0j},$$

and

$$\log\,\{m_{ab}(\boldsymbol{\lambda})m_{rs}(\boldsymbol{\lambda})/m_{as}(\boldsymbol{\lambda})m_{rb}(\boldsymbol{\lambda})\} = \lambda_{ab}.$$

Let $n \to \infty$, $n_{i0} \to \infty$, and $n_{0j} \to \infty$ such that $n_{i0}/n \to \pi_{i0}$ and $n_{0j}/n \to \pi_{0j}$ for all i and j. Under these conditions,

$$h\{n_{ab}; n_{i0}, n_{0j}, \lambda_{ab}\} \sim q\{m_{ij}(\boldsymbol{\lambda})\}\,\exp\,[-\tfrac{1}{2}\,\Sigma_{ij}\,\{n_{ij} - m_{ij}(\boldsymbol{\lambda})\}^2/m_{ij}(\boldsymbol{\lambda})].$$

The expression on the right is the density function of a singular normal multivariate distribution for $(r-1)(s-1)$ variates $\{N_{ab}\}$ in rs dimensions. Singularity arises because the rs frequencies $\{n_{ij}\}$ must satisfy the $r + s - 1$ independent conditions

$$n_{i0} = m_{i0}(\boldsymbol{\lambda}) \quad \text{and} \quad n_{0j} = m_{0j}(\boldsymbol{\lambda}) \quad \text{for all } i \text{ and } j.$$

The proof of asymptotic normality closely follows the lines of the proof in section 4.2. As before, the denominator of $h\{n_{ab}; n_{i0}, n_{0j}, \lambda_{ab}\}$ is eliminated by considering

$$\rho = \log h\{n_{ab}; n_{i0}, n_{0j}, \lambda_{ab}\} - \log h\{m_{ab}(\boldsymbol{\lambda}); n_{i0}, n_{0j}, \lambda_{ab}\}.$$

Stirling's formula is again used to replace the factorials and is followed by an expansion of ρ in powers of $n_{ij} - m_{ij}(\boldsymbol{\lambda})$. At the stage where the various contributions to ρ are collected, we use the property of $\{m_{ij}(\boldsymbol{\lambda})\}$ that

$$\Sigma_{ij} \{n_{ij} - m_{ij}(\boldsymbol{\lambda})\} \log m_{ij}(\boldsymbol{\lambda}) - \Sigma_{ab} \{n_{ab} - m_{ab}(\boldsymbol{\lambda})\}\lambda_{ab} \equiv 0.$$

This identity is verified by checking that the coefficients of $\log m_{ab}(\boldsymbol{\lambda})$, $\log m_{as}(\boldsymbol{\lambda})$, and $\log m_{rs}(\boldsymbol{\lambda})$ are all zero. For example, the coefficient of $\log m_{as}(\boldsymbol{\lambda})$, namely

$$\{n_{as} - m_{as}(\boldsymbol{\lambda})\} + \Sigma_b \{n_{ab} - m_{ab}(\boldsymbol{\lambda})\},$$

is zero because $m_{a0}(\boldsymbol{\lambda}) = n_{a0}$.

We turn next to the moments of $\{N_{ij}\}$. Denote the expectation of N_{ij} by $E(N_{ij}; \boldsymbol{\lambda})$ and the dispersion matrix of $\{N_{ij}\}$ by $\mathfrak{D}\{N_{ij}; \boldsymbol{\lambda}\}$. The form in which the asymptotic distribution of $\{N_{ab}\}$ is expressed makes clear that

$$E(N_{ij}; \boldsymbol{\lambda}) \sim m_{ij}(\boldsymbol{\lambda}).$$

When r and s are small, an approximation to $\mathfrak{D}\{N_{ab}; \boldsymbol{\lambda}\}$ is obtained by expressing the asymptotic density as a function of $\{n_{ab} - m_{ab}(\boldsymbol{\lambda})\}$. An approximation to $\mathfrak{D}\{N_{ij}; \boldsymbol{\lambda}\}$ can then be derived by standard arguments. This method becomes tedious when r and s increase, and a more convenient general procedure is as follows. Consider rs mutually independent variates $\{N_{ij}\}$ such that N_{ij} is normally distributed with mean and variance both $m_{ij}(\boldsymbol{\lambda})$. The variates form a matrix of order $r \times s$. Write $\mathbf{y}_1$ for the vector obtained when the columns of this matrix are adjoined to form a single column. Thus

$$\mathbf{y}_1^T = (N_{11}, N_{21}, \ldots, N_{r1}, N_{12}, N_{22}, \ldots, N_{rs}).$$

Write $\mathbf{y}_2$ for the vector obtained by adjoining N, the $(r-1)$ elements $\{N_{a0}\}$, and the $(s-1)$ elements $\{N_{0b}\}$. Thus

$$\mathbf{y}_2^T = (N, N_{10}, N_{20}, \ldots, N_{r-1,0}, N_{01}, N_{02}, \ldots, N_{0,\,s-1}).$$

Consider the vector $\mathbf{y}$ defined by

$$\mathbf{y}^T = (\mathbf{y}_1^T \,\vdots\, \mathbf{y}_2^T).$$

The dispersion matrix $\mathfrak{D}(\mathbf{y})$ is obtained from $\mathfrak{D}(\mathbf{y}_1)$ by standard arguments. In accordance with the usual notation,

$$\mathfrak{D}\begin{bmatrix} \mathbf{y}_1 \\ \cdots \\ \mathbf{y}_2 \end{bmatrix} = \begin{bmatrix} \mathbf{V}_{11} & \vdots & \mathbf{V}_{12} \\ \cdots & & \cdots \\ \mathbf{V}_{21} & \vdots & \mathbf{V}_{22} \end{bmatrix}.$$

The relevant properties of a normal multivariate distribution (Anderson, 1958, p. 29) imply that the dispersion matrix of $\mathbf{y}_1$ conditional on fixed $\mathbf{y}_2$

is $\mathbf{V}_{11} - \mathbf{V}_{12}\mathbf{V}_{22}^{-1}\mathbf{V}_{21}$. Consequently

$$\mathfrak{D}\{N_{ij};\boldsymbol{\lambda}\} \sim \mathbf{V}_{11} - \mathbf{V}_{12}\mathbf{V}_{22}^{-1}\mathbf{V}_{21}.$$

Finally, we generalize the results of section 4.3. A table of order $r \times s$ provides $(r-1)(s-1)$ cross-product ratios, typified by $N_{ab}N_{rs}/N_{as}N_{rb}$. Write

$$L_{ab} = \log(N_{ab}N_{rs}/N_{as}N_{rb}).$$

The variates $\{L_{ab}\}$ form a matrix of order $(r-1) \times (s-1)$. Denote by $\mathbf{L}$ the vector obtained when the columns of this matrix are adjoined to form a single column. Thus

$$\mathbf{L}^T = (L_{11}, L_{21}, \ldots, L_{r-1,1}, L_{12}, L_{22}, \ldots, L_{r-1,s-1}).$$

Let $\{t_{a0}\}$, $\{t_{0b}\}$, and t be fixed numbers, and write $\mu = \Sigma_{ij}\, \mu_{ij}$. Assign to $\{n_{a0}\}$, $\{n_{0b}\}$, and n the values

$$n_{a0} = [\mu\pi_{a0} + t_{a0}(\mu\pi_{a0})^{\frac{1}{2}}], \quad n_{0b} = [\mu\pi_{0b} + t_{0b}(\mu\pi_{0b})^{\frac{1}{2}}],$$

and $$n = [\mu + t\mu^{\frac{1}{2}}],$$

where again $[x]$ stands for the integral part of x in these equations only. Consider the distribution of $\mu^{\frac{1}{2}}(\mathbf{L} - \boldsymbol{\lambda})$ conditional on $\{N_{a0} = n_{a0}\}$, $\{N_{0b} = n_{0b}\}$, and $N = n$ in the limit as $\mu \to \infty$. Then the conditional distribution converges to a normal multivariate distribution with mean vector zero and dispersion matrix $\Delta\{\pi_{ij}\}$ obtained by treating the variates $\{\mu^{\frac{1}{2}} \log N_{ij}\}$ as mutually independent and such that

$$\operatorname{var}(\mu^{\frac{1}{2}} \log N_{ij}) = 1/\pi_{ij}.$$

The proof follows the same lines as before. An alternative formulation of the result is obtained by the substitution of $m_{ij}(\boldsymbol{\lambda})$ for $\mu\pi_{ij}$.

5.2 Estimation of $\boldsymbol{\lambda}$

The parameters $\{\lambda_{ab}\}$ form a matrix of order $(r-1) \times (s-1)$. Write $\boldsymbol{\lambda}$ for the vector obtained when the columns of this matrix are adjoined to form a single column. Thus

$$\boldsymbol{\lambda}^T = (\lambda_{11}, \lambda_{21}, \ldots, \lambda_{r-1,1}, \lambda_{12}, \lambda_{22}, \ldots, \lambda_{r-1,s-1}).$$

Consider first the point estimation of $\boldsymbol{\lambda}$. The conditional maximum-likelihood estimator is derived from the distribution $h\{n_{ab}; n_{i0}, n_{0j}, \lambda_{ab}\}$ and denoted by $\tilde{\boldsymbol{\lambda}}$. We require the derivatives

$$\partial \log h/\partial\lambda_{ab} = N_{ab} - E(N_{ab}; \boldsymbol{\lambda})$$

and

$$\partial^2 \log h/\partial\lambda_{ab}\partial\lambda_{a'b'} = -\operatorname{cov}(N_{ab}, N_{a'b'}; \boldsymbol{\lambda}).$$

They show that $\tilde{\boldsymbol{\lambda}}$ satisfies the system of equations

$$N_{ab} = E(N_{ab}; \tilde{\boldsymbol{\lambda}}) \quad \text{for all } a \text{ and } b,$$

and that the asymptotic dispersion matrix of $\tilde{\boldsymbol{\lambda}}$ is given by

$$\mathfrak{D}(\tilde{\boldsymbol{\lambda}}) \sim [\mathfrak{D}\{N_{ab}; \boldsymbol{\lambda}\}]^{-1}.$$

The conditional maximum-likelihood estimate is unique whenever it exists (Haberman, 1970, ch. II). In principle, the system of equations for $\tilde{\boldsymbol{\lambda}}$ can be solved by the Newton–Raphson procedure for several unknowns. Denote by $g(\boldsymbol{\lambda})$ the vector of quantities $\{N_{ab} - E(N_{ab}; \boldsymbol{\lambda})\}$. Let $\boldsymbol{\lambda}^{(k)}$ be the kth approximation to $\tilde{\boldsymbol{\lambda}}$. Then

$$\boldsymbol{\lambda}^{(k+1)} = \boldsymbol{\lambda}^{(k)} - [\mathfrak{D}\{N_{ab}; \boldsymbol{\lambda}^{(k)}\}]^{-1} g(\boldsymbol{\lambda}^{(k)}).$$

In practice, the exact calculation of $\mathfrak{D}\{N_{ab}; \boldsymbol{\lambda}^{(k)}\}$ may present considerable problems of enumeration.

The unconditional maximum-likelihood estimator of $\boldsymbol{\lambda}$ is derived from the original distribution of rs mutually independent Poisson variates $\{N_{ij}\}$ with parameters $\{\mu_{ij}\}$ respectively. Denote by $\hat{\boldsymbol{\lambda}}$ the unconditional estimator. The elements of $\hat{\boldsymbol{\lambda}}$ are given by

$$\hat{\boldsymbol{\lambda}}_{ab} = \log(N_{ab} N_{rs} / N_{as} N_{rb}).$$

$E(\hat{\lambda}_{ab})$ does not exist, but the difficulty can be removed by defining the modified estimator

$$\lambda^{+}_{ab} = \log\{(N_{ab} + \tfrac{1}{2})(N_{rs} + \tfrac{1}{2})/(N_{as} + \tfrac{1}{2})(N_{rb} + \tfrac{1}{2})\},$$

which has bias $O(\mu^{-2})$. We can use $\{\hat{\lambda}_{ab}\}$ or $\{\lambda^{+}_{ab}\}$ as initial approximations to $\{\tilde{\lambda}_{ab}\}$ in the Newton–Raphson procedure for calculating the conditional maximum-likelihood estimate.

Consider next the construction of a confidence region for $\boldsymbol{\lambda}$. The question of whether or not to include a particular set of frequencies $\{n_{ab}\}$ can only be answered in somewhat arbitrary terms. Therefore we confine attention to approximate regions inferred from large samples, where the boundary of the region is automatically decided by the procedure used. The corresponding methods of section 4.4 are generalized as follows. Suppose that the conditions are such that the frequencies $\{N_{ab}\}$ have an asymptotic normal multivariate distribution. The asymptotic density function of $\{N_{ab}\}$ is proportional to $\exp(-\frac{1}{2} X^2)$, where

$$X^2 = \Sigma_{ij} \{N_{ij} - m_{ij}(\boldsymbol{\lambda})\}^2 / m_{ij}(\boldsymbol{\lambda}).$$

A further appeal to normal multivariate theory (Anderson, 1958, p. 54) shows that the asymptotic distribution of X^2 is $\chi^2_{(r-1)(s-1)}$. Let k^2_ϵ be the upper 100ϵ percent point of $\chi^2_{(r-1)(s-1)}$. Then the set of vectors $\boldsymbol{\lambda}$ such that

$X^2 \leqslant k_\epsilon^2$ is a confidence region for $\boldsymbol{\lambda}$ with confidence coefficient approximately $1 - \epsilon$. The problems of enumeration involved here escalate rapidly when $(r-1)(s-1)$ increases.

A more convenient procedure for constructing a confidence region is based on the asymptotic conditional distribution of $\mu^{\frac{1}{2}}(\hat{\lambda} - \lambda)$ derived in the previous section, where $\hat{\boldsymbol{\lambda}}$ was denoted by $\mathbf{L}$. Consider the variate

$$T^2 = \mu^{(r-1)(s-1)}(\hat{\boldsymbol{\lambda}} - \boldsymbol{\lambda})^T [\Delta\{\pi_{ij}\}]^{-1} (\hat{\boldsymbol{\lambda}} - \boldsymbol{\lambda}).$$

Then T^2 converges in distribution to $\chi^2_{(r-1)(s-1)}$. Since μ/N converges in probability to 1 and N_{ij}/N converges in probability to π_{ij}, a theorem of Cramér (1946, sect. 20.6) ensures that the same limiting distribution holds for

$$U^2 = N^{(r-1)(s-1)} (\hat{\boldsymbol{\lambda}} - \boldsymbol{\lambda})^T [\Delta\{N_{ij}/N\}]^{-1} (\hat{\boldsymbol{\lambda}} - \boldsymbol{\lambda}).$$

Each element of Δ is either zero or a linear combination of $\{\pi_{ij}^{-1}\}$, whence

$$U^2 = (\hat{\boldsymbol{\lambda}} - \boldsymbol{\lambda})^T [\Delta\{N_{ij}\}]^{-1} (\hat{\boldsymbol{\lambda}} - \boldsymbol{\lambda}).$$

For the reasons given in section 4.4, we use $\boldsymbol{\lambda}^+$ in place of $\hat{\boldsymbol{\lambda}}$ and replace N_{ij} by $N_{ij} + \frac{1}{2}$. An asymptotically equivalent statistic is therefore

$$Z^2 = (\boldsymbol{\lambda}^+ - \boldsymbol{\lambda})^T [\Delta\{N_{ij} + \tfrac{1}{2}\}]^{-1} (\boldsymbol{\lambda}^+ - \boldsymbol{\lambda}).$$

Then the set of vectors $\boldsymbol{\lambda}$ such that $Z^2 \leqslant k_\epsilon^2$ is a confidence region for $\boldsymbol{\lambda}$ with confidence coefficient approximately $1 - \epsilon$. We can infer from the discussion of the logit method in section 4.4 that the confidence region based on Z^2 is a less accurate approximation than the region based on X^2. Either method leads to simultaneous intervals of the type described by Scheffé (1959, sect. 3.5); the details are given by Goodman (1964). Another procedure of making multiple comparisons, based on the likelihood-ratio statistic, is described by Gabriel (1966).

Example 5.1 The data are given in Table 3.3. We proceed to derive a confidence region for λ_1 and λ_2, defined by

$$\lambda_1 = \log(p_{11}p_{22}/p_{12}p_{21}) \quad \text{and} \quad \lambda_2 = \log(p_{11}p_{23}/p_{13}p_{21}).$$

The modified unconditional estimates are as follows.

$$\lambda_1^+ = \log\{(19{\cdot}5)(560{\cdot}5)/(29{\cdot}5)(497{\cdot}5)\} = -0{\cdot}294\,742.$$

$$\lambda_2^+ = \log\{(19{\cdot}5)(269{\cdot}5)/(24{\cdot}5)(497{\cdot}5)\} = -0{\cdot}841\,286.$$

Their estimated variances and covariance form the matrix $\Delta\{n_{ij} + \frac{1}{2}\}$:

$$\text{est var}(\lambda_1^+) = (19{\cdot}5)^{-1} + (560{\cdot}5)^{-1} + (29{\cdot}5)^{-1} + (497{\cdot}5)^{-1} = 0{\cdot}088\,974,$$

$$\text{est var}(\lambda_2^+) = (19{\cdot}5)^{-1} + (269{\cdot}5)^{-1} + (24{\cdot}5)^{-1} + (497{\cdot}5)^{-1} = 0{\cdot}097\,819,$$

$$\text{est cov}(\lambda_1^+, \lambda_2^+) = (19{\cdot}5)^{-1} + (497{\cdot}5)^{-1} = 0{\cdot}053\,292.$$

The upper 5 percent point of χ_2^2 is 5·991. A confidence region for (λ_1, λ_2) with confidence coefficient 0·95 consists of all pairs (λ_1, λ_2) such that

$$(\boldsymbol{\lambda}^+ - \boldsymbol{\lambda})^T\,[\Delta\{n_{ij} + \tfrac{1}{2}\}]^{-1}\,(\boldsymbol{\lambda}^+ - \boldsymbol{\lambda}) \leqslant 5{\cdot}991.$$

The resulting region is shown in Fig. 5.1. There is a significant logit increase for the second category of enlarged tonsils (+++), but not for the first (++).

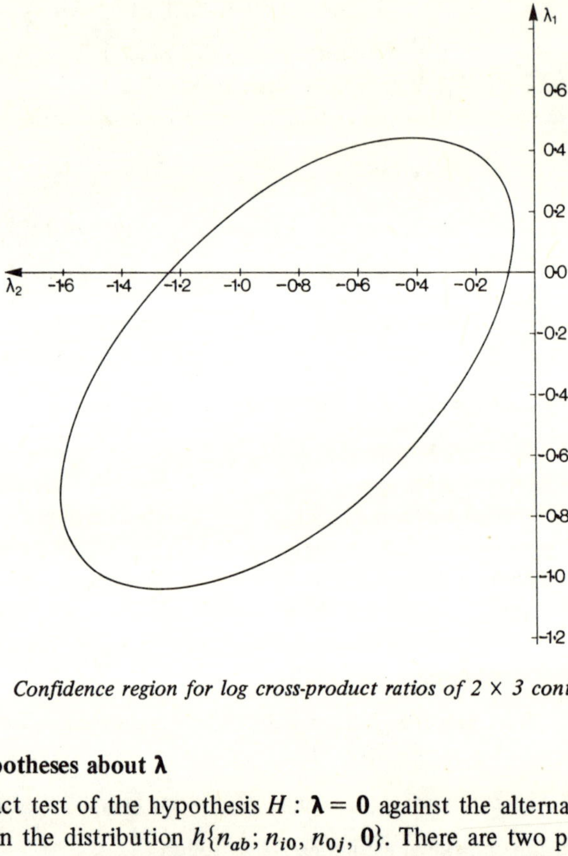

Fig. 5.1 *Confidence region for log cross-product ratios of 2 × 3 contingency table*

5.3 Hypotheses about $\boldsymbol{\lambda}$

An exact test of the hypothesis $H : \boldsymbol{\lambda} = \mathbf{0}$ against the alternative $K : \boldsymbol{\lambda} \neq \mathbf{0}$ is based on the distribution $h\{n_{ab}; n_{i0}, n_{0j}, \mathbf{0}\}$. There are two problems. First, we have to decide which tables are more significant than the one observed. We can rank tables either by their individual probabilities, or by the total probabilities of sets obtained from a specified table by permuting rows and columns, or by the value of some test statistic, and the rankings will generally differ (Healy, 1969). Secondly, we have to calculate the total probability of tables which are at least as significant as the one observed. The calculations involved increase rapidly with r and s (Freeman and Halton, 1951). Indeed, the difficulties are such as to direct attention towards methods of testing H against K which are strictly appropriate only when the marginal totals are large.

The theory of testing a simple hypothesis against a composite alternative was reviewed in section 2.3. We proceed to evaluate the statistic S_3 defined there. The elements of $n^{\frac{1}{2}}\mathbf{y}_0$ are typified by

$$(\partial \log h/\partial\lambda_{ab})_0 = N_{ab} - E_0(N_{ab}).$$

Similarly, a typical element of $n\mathbf{K}_0$ is

$$-E(\partial^2 \log h/\partial\lambda_{ab}\,\partial\lambda_{a'b'})_0 = \mathrm{cov}_0(N_{ab}, N_{a'b'}).$$

The expression $\mathbf{y}_0^T\mathbf{K}_0^{-1}\mathbf{y}_0$ does not involve the frequencies in the last row or column of the contingency table, and a more symmetrical expression is obtained as follows. Denote by $\mathbf{x}_0$ the vector with elements $\{N_{ij} - E_0(N_{ij})\}$, and let $\mathbf{J}_0$ be the matrix with elements $\mathrm{cov}_0(N_{ij}, N_{i'j'})$. $\mathbf{J}_0$ is singular, but there exists a generalized inverse $\mathbf{J}_0^{g_1}$ such that

$$\mathbf{J}_0\mathbf{J}_0^{g_1}\mathbf{J}_0 = \mathbf{J}_0.$$

For the reasons already given in section 2.3, we have the alternative expression

$$S_3 = \mathbf{x}_0^T\mathbf{J}_0^{g_1}\mathbf{x}_0.$$

The moments of $\{N_{ij}\}$ required here are

$$E_0(N_{ij}) = n_{i0}n_{0j}/n,$$

and $\quad \mathrm{cov}_0(N_{ij}, N_{i'j'}) = n_{i0}(\delta_{ii'}n - n_{i'0})n_{0j}(\delta_{jj'}n - n_{0j'})/n^2(n-1).$

Define vectors and matrices as follows:

$$\boldsymbol{\rho}^T = (n_{10}, n_{20}, \ldots, n_{r0}), \quad \boldsymbol{\kappa}^T = (n_{01}, n_{02}, \ldots, n_{0s}),$$

$$\mathbf{d}_\rho = \mathrm{diag}(n_{10}, n_{20}, \ldots, n_{r0}), \quad \mathbf{d}_\kappa = \mathrm{diag}(n_{01}, n_{02}, \ldots, n_{0s}),$$

$$\mathbf{A} = (n\mathbf{I} - \rho\mathbf{1}^T)\mathbf{d}_\rho, \quad \text{and} \quad \mathbf{B} = (n\mathbf{I} - \kappa\mathbf{1}^T)\mathbf{d}_\kappa.$$

In terms of these quantities,

$$n^2(n-1)\mathbf{J}_0 = \mathbf{A}\otimes\mathbf{B},$$

where $\otimes$ denotes a direct product. We now show that the condition $\mathbf{J}_0\mathbf{J}_0^{g_1}\mathbf{J}_0 = \mathbf{J}_0$ is satisfied by

$$\mathbf{J}_0^{g_1} = (n-1)\mathbf{d}_\rho^{-1}\otimes\mathbf{d}_\kappa^{-1}.$$

Since $(\mathbf{L}\otimes\mathbf{M})(\mathbf{P}\otimes\mathbf{Q}) = (\mathbf{LP})\otimes(\mathbf{MQ})$, it suffices to verify that the condition holds for the first term in the representation of each matrix as a direct product. In fact,

$$\mathbf{A}\mathbf{d}_\rho^{-1}\mathbf{A} = (n\mathbf{I} - \rho\mathbf{1}^T)(n\mathbf{I} - \rho\mathbf{1}^T)\mathbf{d}_\rho,$$

which reduces to $n\mathrm{A}$ on using $\mathbf{1}^T\boldsymbol{\rho} = n$. Consequently

$$S_3 = (n-1)\,\Sigma_{ij}\,(N_{ij} - n_{io}n_{oj}/n)^2/n_{io}n_{oj}.$$

On the other hand,

$$X^2 = n\,\Sigma_{ij}\,(N_{ij} - n_{io}n_{oj}/n)^2/n_{io}n_{oj},$$

using the result that

$$m_{ij}(\mathbf{0}) = n_{io}n_{oj}/n.$$

We have therefore shown that

$$S_3 = (n-1)X^2/n.$$

The same result is noted by Birch (1965), and proved for $2 \times s$ tables by Gart (1972). Studies of the exact distribution of X^2 when $\boldsymbol{\lambda} = 0$ indicate that agreement with the asymptotic results is good even when the frequencies are quite small (Haynam and Leone, 1965; Lewontin and Felsenstein, 1965; Craddock and Flood, 1970).

We conclude with brief remarks on the general problem of testing $H : \boldsymbol{\lambda} = \boldsymbol{\lambda}_0$ against $K : \boldsymbol{\lambda} \neq \boldsymbol{\lambda}_0$. The main point to establish is that $m_{ij}(\boldsymbol{\lambda}_0)$ is the maximum-likelihood estimate of μ_{ij} under H. This is proved by introducing Lagrangian multipliers $\{\xi_{ab}\}$, and differentiating

$$\Sigma_{ij}\,n_{ij}\log\mu_{ij} - \Sigma_{ij}\,\mu_{ij} - \Sigma_{ab}\,\xi_{ab}(\log\mu_{ab} - \log\mu_{as} - \log\mu_{rb} + \log\mu_{rs})$$

with respect to $\{\mu_{ij}\}$. There are different forms of derivative for μ_{ab}, μ_{as}, and μ_{rs}. They lead respectively to

$$n_{ab} = \hat{\mu}_{ab} + \xi_{ab}, \quad n_{as} = \hat{\mu}_{as} - \Sigma_b\,\xi_{ab},$$

and

$$n_{rs} = \hat{\mu}_{rs} + \Sigma_{ab}\,\xi_{ab}.$$

Summation over i and j gives

$$\hat{\mu}_{io} = n_{io} \quad \text{and} \quad \hat{\mu}_{oj} = n_{oj}.$$

The estimates are therefore obtained from the marginal totals by iterative scaling based on the specified values of $\{\lambda_{ab}\}$, whence

$$\hat{\mu}_{ij} = m_{ij}(\boldsymbol{\lambda}_0).$$

A further generalization of the arguments in section 4.5 leads to the test criteria

$$X^2 = \Sigma_{ij}\,\{N_{ij} - m_{ij}(\boldsymbol{\lambda}_0)\}^2/m_{ij}(\boldsymbol{\lambda}_0)$$

and

$$Y^2 = 2\,\Sigma_{ij}\,N_{ij}\log\{N_{ij}/m_{ij}(\boldsymbol{\lambda}_0)\}.$$

Each is distributed asymptotically as $\chi^2_{(r-1)(s-1)}$ when H is true. The asymptotic power function of X^2 has been derived from the general results of

Mitra (1958) by Meng and Chapman (1966).

Example 5.2 The data are given in Table 3.4. We shall test the hypothesis that teacher's rating and homework conditions are independent. Two large-sample test criteria are evaluated for comparison:

$$X^2 = 9{\cdot}093, \quad Y^2 = 9{\cdot}037.$$

Each is distributed as χ_8^2 under independence, and gives no evidence for rejection.

5.4 Partitions of chi-square

The likelihood-ratio statistic Y^2 can be expressed as the sum of $(r-1)\times(s-1)$ components which are mutually independent and asymptotically χ_1^2 when $\boldsymbol{\lambda} = \mathbf{0}$. There are many ways in which the components can be chosen or grouped. The partition of Y^2 selected for analysis will depend on those aspects of the table considered which have a meaningful interpretation when a set of components is declared significant. Similar partitions obtain for Pearson's statistic X^2, in connection with which the methods to be described were first proposed (Lancaster, 1949). However, the sum of the components of X^2 is equal to the whole only asymptotically.

The justification for these statements is as follows. Consider a 2×3 table, and write

$$n_{11} + n_{12} = a_{12}, \quad n_{21} + n_{22} = a_{22}, \quad n_{01} + n_{02} = a_{02}.$$

When $\boldsymbol{\lambda} = \mathbf{0}$, the distribution of $\{N_{ij}\}$, conditional on the observed values of $\{N_{io}\}$ and $\{N_{oj}\}$, admits the product representation

$$\frac{n_{10}!\,n_{20}!\,n_{01}!\,n_{02}!\,n_{03}!}{n!\,\Pi_{ij}\,n_{ij}!} = \left\{\frac{a_{12}!\,a_{22}!\,n_{01}!\,n_{02}!}{a_{02}!\,n_{11}!\,n_{12}!\,n_{21}!\,n_{22}!}\right\}\left\{\frac{n_{10}!\,n_{20}!\,a_{02}!\,n_{03}!}{n!\,a_{12}!\,a_{22}!\,n_{13}!\,n_{23}!}\right\}.$$

The terms in brackets on the right correspond respectively to the 2×2 tables shown below. Each term is the distribution for the entries in the table, conditional on the observed marginal totals and on the assumption that the corresponding value of λ is zero. A consequence of the product representation is that any function of the entries in the first table is independent of any function of the entries in the second.

n_{11}	n_{12}	a_{12}
n_{21}	n_{22}	a_{22}
n_{01}	n_{02}	a_{02}

a_{12}	n_{13}	n_{10}
a_{22}	n_{23}	n_{20}
a_{02}	n_{03}	n

The observed value of the likelihood-ratio statistic for testing the hypothesis that $\boldsymbol{\lambda} = 0$ in a table of order $r \times s$ is

$$2\,\Sigma_{ij}\, n_{ij} \log (n n_{ij}/n_{i0} n_{0j})$$
$$= 2\,(\Sigma_{ij}\, n_{ij} \log n_{ij} - \Sigma_i\, n_{i0} \log n_{i0} - \Sigma_j\, n_{0j} \log n_{0j} + n \log n).$$

This expression shows that the statistic is calculated from the values of $2f \log f$, where f denotes frequency, with a positive sign for $\{n_{ij}\}$ and n, and a negative sign for $\{n_{i0}\}$ and $\{n_{0j}\}$. Denote by Y_1^2 and Y_2^2 the likelihood-ratio statistics for the constituent 2×2 tables of the 2×3 table under consideration. The patterns of signs for $2f \log f$ in the constituent tables are shown below.

$$\begin{array}{cc|c} + & + & - \\ + & + & - \\ \hline - & - & + \end{array} \qquad \begin{array}{cc|c} + & + & - \\ + & + & - \\ \hline - & - & + \end{array}$$

Since the last column of the first table of frequencies is identical with the first column of the second table,

$$Y^2 = Y_1^2 + Y_2^2.$$

The distribution theory shows that Y_1^2 and Y_2^2 are independent and asymptotically χ_1^2 when $\boldsymbol{\lambda} = \mathbf{0}$. We have therefore proved the assertion at the beginning of this section for a table of order 2×3. A table of order $2 \times s$ can be similarly divided into a set of $(s - 1)$ constituent 2×2 tables, and a table of order $r \times s$ gives rise to $(r - 1)$ such sets. The same method of proof then establishes that the $(r - 1)(s - 1)$ values of Y^2 from individual 2×2 tables form an exact partition of the overall Y^2 and are mutually independent and asymptotically χ_1^2 when $\boldsymbol{\lambda} = \mathbf{0}$.

We can instead calculate X^2 from each of the $(r - 1)(s - 1)$ constituent 2×2 tables. The corresponding values are mutually independent and asymptotically χ_1^2 when $\boldsymbol{\lambda} = \mathbf{0}$, but their total is not exactly the overall X^2. However, the partition is asymptotically exact because X^2 and Y^2 are asymptotically equivalent.

Example 5.3 The relationship between nasal carrier rate for *Streptococcus pyogenes* and size of tonsils among children aged 0–15 years is shown in Table 3.3. We form 2×2 tables from the right as follows:

	Not enlarged	Enlarged	++	+++
Carriers	19	53	29	24
Non-carriers	497	829	560	269

The corresponding values of Y^2 are

$$Y_1^2 = 3{\cdot}784, \quad Y_2^2 = 3{\cdot}537.$$

Thus the differences between carrier rates are associated with both the existence and extent of enlargement.

5.5 Symmetric tables

Suppose that each unit of a population has two characteristics which are classified by the same method. An example is given by the data on women's unaided distance vision in each eye (Table 3.5). In such cases, the contingency table has order $r \times r$, and two hypotheses naturally present themselves. First, there is the hypothesis of *symmetry* defined by

$$H_1 : p_{ij} = p_{ji} \quad \text{for all } i, j \text{ with } i < j.$$

Secondly, there is the hypothesis of *marginal homogeneity*, defined by

$$H_2 : p_{io} = p_{oi} \quad \text{for all } i.$$

When $r = 2$, H_1 and H_2 are equivalent, but otherwise their relationship can be expressed as follows. The hypothesis of *quasi-symmetry*, due to Caussinus (1965), is given by

$$H_3 : \lambda_{ab} = \lambda_{ba} \quad \text{for all } a, b \text{ with } a < b.$$

Symmetry evidently implies both marginal homogeneity and quasi-symmetry. On the other hand, the table is uniquely determined by the marginal probabilities and cross-product ratios, so that marginal homogeneity and quasi-symmetry together imply symmetry. The elements in the main diagonal of the table carry no information about any of these hypotheses, so that all the analyses which follow can be carried out on the truncated table obtained by deleting them.

We now consider the estimation of expected frequencies for the three hypotheses. Under the hypothesis of symmetry, the probabilities are subject to the condition

$$\Sigma_i \, p_{ii} + 2\, \Sigma_{i<j} \, p_{ij} = 1.$$

Denote the maximum-likelihood estimator of np_{ij} by m_{1ij}. Then

$$m_{1ij} = (N_{ij} + N_{ji})/2.$$

Thus H_1 can be tested by referring either

$$X_1^2 = \Sigma_{i<j} \, (N_{ij} - N_{ji})^2/(N_{ij} + N_{ji})$$

or

$$Y_1^2 = 2\, \Sigma_{i \neq j} \, N_{ij} \log \{2N_{ij}/(N_{ij} + N_{ji})\}$$

to a table of $\chi^2_{r(r-1)/2}$ (Bowker, 1948).

Under the hypothesis of marginal homogeneity, denote the maximum-likelihood estimator of np_{ij} by m_{2ij}. The estimators are obtained by maximizing

$$\Sigma_{ij} N_{ij} \log p_{ij} - n\, \Sigma_i\, \alpha_i\, (p_{io} - p_{oi}) - \beta \Sigma_{ij}\, p_{ij}$$

with respect to $\{p_{ij}\}$, where $\{\alpha_i\}$ and β are Lagrangian multipliers. This leads to the equations

$$m_{2io} = m_{2oi}, \quad \Sigma_{ij}\, m_{2ij} = n,$$

and

$$N_{ij}/m_{2ij} = 1 + \alpha_i - \alpha_j.$$

Clearly $m_{2ii} = N_{ii}$. The other expected frequencies are calculated as follows. We note that the $\{\alpha_i\}$ appear only as differences $\{\alpha_i - \alpha_j\}$, and therefore take $\alpha_r = 0$ throughout. Suppose that a_i is an approximation to α_i. According to the Newton–Raphson method, a better approximation is $a_i + d_i$, where the $\{d_i\}$ are found by substituting

$$m_{2ij} = n_{ij}\{1/(1 + a_i - a_j) - (d_i - d_j)/(1 + a_i - a_j)^2\}$$

in the equations $m_{2io} = m_{2oi}$. There are $(r - 1)$ independent equations for the $(r - 1)$ unknowns $\{d_i\}$, where $d_r = 0$. The solution gives a second approximation for $\{\alpha_i\}$. We begin the process with $a_i = 0$ for all i and continue until a satisfactory agreement of the two sets of marginal totals is reached. A test of H_2 is made by referring either

$$X_2^2 = \Sigma_{i \neq j}\, (N_{ij} - m_{2ij})^2/m_{2ij}$$

or

$$Y_2^2 = 2\, \Sigma_{i \neq j}\, N_{ij} \log\, (N_{ij}/m_{2ij})$$

to a table of χ^2_{r-1}. The approach is equivalent to one made by Wedderburn (1974).

We conclude with the hypothesis of quasi-symmetry. Denote the maximum-likelihood estimator of np_{ij} by m_{3ij}. Proceeding as before, we obtain the equations

$$m_{3io} = N_{io}, \quad m_{3oj} = N_{oj},$$

$$m_{3ab} m_{3rr}/m_{3ar} m_{3rb} = m_{3ba} m_{3rr}/m_{3br} m_{3ra},$$

and

$$m_{3ij} + m_{3ji} = N_{ij} + N_{ji} \quad (i \leqslant j).$$

Once again, the observed diagonal elements are the corresponding expected frequencies. A convergent iterative scheme for the calculation of the other expected frequencies is as follows (Ireland, Ku and Kullback, 1969). We begin with the expected frequencies under the hypothesis of symmetry. They satisfy the last two conditions but not the first two. The procedure continues by iterative scaling on the row and column totals, alternating with iterative scaling on the totals $\{m_{3ij} + m_{3ji}\}$. These operations preserve $\lambda_{ab} = \lambda_{ba}$ throughout, and ultimately lead to the correct values for all three sets of totals. Finally, H_3 is tested by referring either

$$X_3^2 = \Sigma_{i \neq j} (N_{ij} - m_{3ij})^2 / m_{3ij}$$

or

$$Y_3^2 = 2 \Sigma_{i \neq j} N_{ij} \log (N_{ij}/m_{3ij}),$$

to a table of $\chi^2_{(r-1)(r-2)/2}$.

The fact that symmetry is equivalent to the combination of marginal homogeneity and quasi-symmetry implies that $Y_1^2 - Y_3^2$ is the likelihood-ratio statistic for testing marginal homogeneity on the assumption that quasi-symmetry holds.

Example 5.4 For the data in Table 3.5, the results of testing the hypotheses described above are as follows:

(i) Symmetry. $Y_1^2 = 19{\cdot}250$, whereas the upper 1 percent point of χ_6^2 is 16·81.

(ii) Marginal homogeneity. $Y_2^2 = 11{\cdot}986$, whereas the upper 1 percent point of χ_3^2 is 11·34. The expected frequencies are shown in Table 5.1.

(iii) Quasi-symmetry. $Y_3^2 = 7{\cdot}274$, whereas the upper 10 percent point of χ_3^2 is 7·78. The expected frequencies are shown in Table 5.2.

(iv) Marginal homogeneity assuming quasi-symmetry. $Y_1^2 - Y_3^2 = 11{\cdot}976$ which again is highly significant.

TABLE 5.1 *Unaided distance vision of 7477 women aged 30–39: expected frequencies for hypothesis of marginal homogeneity*

Grade of right eye	Grade of left eye: Highest	Second	Third	Lowest	Total
Highest	1520·000	252·482	111·843	56·966	1941·291
Second	247·237	1512·000	409·418	70·585	2239·240
Third	131·269	383·133	1772·000	195·258	2481·660
Lowest	42·785	91·625	188·399	492·000	814·809
Total	1941·291	2239·240	2481·660	814·809	7477·000

TABLE 5.2 *Unaided distance vision of 7477 women aged 30–39: expected frequencies for hypothesis of quasi-symmetry*

Grade of right eye	Grade of left eye: Highest	Second	Third	Lowest	Total
Highest	1520·000	263·380	133·584	59·036	1976·000
Second	236·620	1512·000	418·986	88·394	2256·000
Third	107·416	375·014	1772·000	201·570	2456·000
Lowest	42·964	71·606	182·430	492·000	789·000
Total	1907·000	2222·000	2507·000	841·000	7477·000

We conclude that the table is quasi-symmetrical but has no other form of symmetry. An examination of residuals suggests that vision in the right eye is better than in the left.

5.6 Incomplete tables

Consider a contingency table which corresponds to the distribution of a bivariate response (R, S). We have assumed so far that p_{ij} is positive for all i and j. A table with this property is described as *complete*. By contrast, the table is *incomplete* or *truncated* when some of the cell probabilities are zero. An example is given by the data on fingerprints classified by small loops and whorls (Table 3.6). We say that a table is *separable* if the rows and columns can be permuted so that all the cells with positive probabilities fall into non-overlapping blocks down the main diagonal. Each block of a separable table can be treated either by the methods already described or by those in the present section. Again, if $p_{ij} = 0$ throughout a whole row or column, then no information is available about the corresponding category, which can be omitted. We therefore confine attention to inseparable incomplete tables from which all null categories have been removed.

A cell where $p_{ij} = 0$ is described as *empty*. For the cells which are not empty, we can express $\log p_{ij}$ in the form

$$\log p_{ij} = \kappa + \theta_i + \phi_j + \lambda_{ij}.$$

The problem of estimability for the parameters on the right can be solved by the introduction of linear restrictions on them. There are two ways in which the restrictions can be formulated. The first follows the approach adopted previously. Suppose that without loss of generality the rows and columns are labelled such that cell (r, s) is not empty, and denote by e the number of empty cells. We then take $\theta_r = 0$, $\phi_s = 0$, and retain exactly g parameters $\{\lambda_{ij}\}$, where

$$g = \max\{0, (r-1)(s-1)-e\}.$$

Alternatively, an analogy with the analysis of variance suggests

$$\Sigma_i\, \theta_i = \Sigma_j\, \phi_j = 0,$$

and
$$\Sigma_i\, \delta_{ij}\lambda_{ij} = \Sigma_j\, \delta_{ij}\lambda_{ij} = 0,$$

where
$$\delta_{ij} = \begin{cases} 1 & \text{when cell } (i, j) \text{ is not empty,} \\ 0 & \text{otherwise.} \end{cases}$$

Suppose that $(r-1)(s-1) > e$. We say that the responses R and S are *quasi-independent* if $\lambda_{ij} = 0$ for all non-empty cells. In other words,

$$p_{ij} = \begin{cases} \alpha_i\beta_j & \text{for non-empty cells,} \\ 0 & \text{elsewhere.} \end{cases}$$

When $(r-1)(s-1) \leqslant e$, this hypothesis is satisfied automatically.

The expected frequencies $\{m_{ij}\}$ for the hypothesis of quasi-independence can always be calculated by the iterative scaling procedure, provided that the observed marginal totals are positive, and the table of observed frequencies is itself inseparable (Fienberg, 1970a). We begin with a table where the frequency is 1 for each non-empty cell and 0 otherwise, and continue by scaling the row and column totals alternately until the observed values are achieved to a sufficient degree of accuracy. The test statistic is either

$$X^2 = \Sigma\,(N_{ij} - m_{ij})^2/m_{ij} \quad \text{or} \quad Y^2 = 2\,\Sigma\,N_{ij}\log(N_{ij}/m_{ij}).$$

The concept of quasi-independence was proposed by White (1963) and Goodman (1963a), although that term did not appear until later.

Consider a contingency table of order $r \times r$ such that $p_{ij} = 0$ when $i + j > r$. The data in Table 3.6 can be rearranged in this form by defining

$$R = \text{Number of whorls} + 1,$$

$$S = \text{Number of small loops} + 1.$$

In view of the constraint $R + S \leqslant r + 1$, the responses R and S must necessarily be dependent. A study of the question whether dependence exists which cannot be attributed to the constraint led Darroch (1971) to put forward the concept of *F-independence*, according to which

$$p_{ij} = \begin{cases} \alpha_i\beta_j\gamma_{r+1-i-j} & \text{for non-empty cells} \\ 0 & \text{otherwise.} \end{cases}$$

The theory of incomplete tables is studied by Caussinus (1965) and the methodology is reviewed by Fienberg (1972).

Example 5.5 The data in Table 3.6 were analysed by Waite (1915) in a noteworthy anticipation of the ideas presented above. At the suggestion of Karl Pearson, he calculated the expected frequencies for the hypothesis of quasi-independence which are given in Table 5.3. The value of X^2 is 399·8 on 10 degrees of freedom, and so this hypothesis is decisively rejected. On the other hand, the data are consistent with the hypothesis of F-independence (Darroch and Ratcliff, 1973).

TABLE 5.3 *Fingerprints of the right hand classified by the numbers of whorls and small loops: expected frequencies for hypothesis of quasi-independence*

Whorls	Small loops 0	1	2	3	4	5	Total
0	200·6	167·4	166·6	150·3	131·1	45·0	861·0
1	122·2	101·9	101·4	91·6	79·9		497·0
2	85·5	71·4	71·0	64·1			292·0
3	63·8	53·2	53·0				170·0
4	70·9	59·1					130·0
5	50·0						50·0
Total	593·0	453·0	392·0	306·0	211·0	45·0	2000·0

6 MODELS AND METHODS

6.1 Statistical models

Methods of classification which distinguish between the categories of a response or the levels of a factor may be qualitative, ordered, or quantitative. The procedures of inference explained in previous chapters are basically intended for qualitative classifications because no assumptions are made about parameters such as $\{\theta_a\}$ or $\{\lambda_{ab}\}$. Classifications which are ordered or quantitative implicitly contain further information about the parameters which can be used to make inferences more precise. We proceed to outline some ways of extracting the information.

Suppose that the levels of a factor $\mathfrak{F}$ are distinguished by a single explanatory variable, such that the value x_j is attached to the jth level of $\mathfrak{F}(j = 1, 2, \ldots, s)$. A response R is observed at each level of $\mathfrak{F}$. The simplest assumption consistent with the approach of previous chapters is a regression model expressed by

$$\lambda_{ab} = \beta_a(x_b - x_s) \quad (a = 1, 2, \ldots, r-1;\ b = 1, 2, \ldots, s-1).$$

Denote by $\boldsymbol{\beta}$ the vector of parameters $\{\beta_a\}$. The problem is to construct estimation and test procedures for $\boldsymbol{\beta}$. Methods of analysis for the case $r = 2$ are discussed in the following section. More complex models, which may involve polynomial relationships or multiple regression, are considered under the general formulation discussed in section 6.4.

Example 6.1 The data quoted in Table 3.2 classify the period of survival of *Tribolium castaneum* into three categories for each of six levels of the deposit of 0·1% w/v γ-benzene hexachloride. Here x_j is the deposit measured in mg/10 cm^2, or a monotonic function of deposit such as the logarithm.

Suppose that there is a natural ordering among the categories of a response R. We can introduce an explanatory variable by attaching a score c_i to the ith category, where

$$c_1 < c_2 < \ldots < c_r.$$

Usually the values of $\{c_i\}$ are equally spaced, but other spacings may be more appropriate in special circumstances. When a bivariate response is observed, where each response is ordered, we can introduce scores $\{c_i\}$ for R and $\{d_j\}$ for S. The basic assumption consistent with our line of approach is then

$$\lambda_{ab} = \beta(c_a - c_r)(d_b - d_s) \quad (a = 1, 2, \ldots, r-1;\ b = 1, 2, \ldots, s-1).$$

This model was proposed by Birch (1965) in a development of work by Yates (1948), Williams (1952), and Mantel (1963). If both $\{c_i\}$ and $\{d_j\}$ are equally spaced, then the model implies that each combination of a pair of successive rows with a pair of successive columns has the same cross-product ratio. The problem is to make inferences about β.

Example 6.2 The methods of classification in Table 3.4 are each ordered. Here we assign scores 1, 0, −1 respectively to the categories A, B, C of teacher's rating, and scores 2, 1, 0, −1, −2 respectively to the categories A, B, C, D, E of homework conditions.

In the remainder of this section, we review other procedures which may be used for the analysis of a contingency table when at least one of the methods of classification is quantitative or ordered. Suppose first that a binary response R is observed at each level of a quantitative factor $\mathfrak{F}$, where the levels of $\mathfrak{F}$ are distinguished by a single explanatory variable. The value x_j attached to the jth level often measures a stimulus applied to units in the corresponding population. In such experiments, each unit of the population is assumed to have a *tolerance* t, so that a stimulus of at least t leads to $R = 1$, but otherwise $R = 2$. The corresponding variate T has the property that

$$p_{1j} = p(T \leqslant x_j) \quad \text{and} \quad p_{2j} = p(T > x_j).$$

We complete the statistical model by specifying the distribution of tolerances. When x_j ranges over $(-\infty, \infty)$, a common assumption is that T is normally distributed with mean $-\alpha/\theta$ and standard deviation $1/\theta$. Thus

$$p_{1j} = \Phi(\alpha + \theta x_j),$$

where $\Phi(.)$ is the distribution function of $\mathfrak{N}(0, 1)$. A large amount of methodology is based on this assumption, and the details are given by Finney (1971). Even when the sample sizes are not small, maximum-likelihood estimation can lead to biases which are appreciable and asymptotic variances which are considerably in error (Sowden, 1971, 1972).

When a response has r ordered categories, the range of T is divided into a set of r mutually exclusive and exhaustive intervals, and p_{ij} is the probability that T falls in the ith interval when the stimulus is x_j (Ashford, 1959). Thus the normal tolerance distribution assumed above leads to

$$p_{ij} = \Phi(\alpha_i + \theta x_j) - \Phi(\alpha_{i-1} + \theta x_j),$$

where $\quad \alpha_0 = -\infty, \quad \alpha_r = \infty, \quad$ and $\quad \alpha_1 < \alpha_2 < \ldots < \alpha_{r-1}.$

Similar extensions apply to other tolerance distributions such as the logistic (Cox, 1970a, sect. 7.4).

An objection to the use of scores for the categories of an ordered classification is that the values are usually quite arbitrary. Some ways of meeting this criticism are as follows.

(a) The scores are estimated from the observed frequencies. When an ordered response is observed in several populations, we assume that categories are formed by dividing an underlying continuous scale of measurement into intervals and then estimate the points of subdivision (Snell, 1964). A simpler method is to assign percentile scores estimated from the marginal totals (Williams and Grizzle, 1972). When two ordered responses are observed simultaneously, scores $\{x_i\}$ are assigned to the categories of the first response and scores $\{y_j\}$ to those of the second in such a way as to maximize the correlation coefficient

$$c = \Sigma_{ij} N_{ij} x_i y_j / n,$$

subject to the conditions

$$\Sigma_i N_{i0} x_i = 0, \qquad \Sigma_j N_{0j} y_j = 0,$$

$$\Sigma_i N_{i0} x_i^2 = n, \quad \text{and} \quad \Sigma_j N_{0j} y_j^2 = n.$$

This is a procedure of canonical analysis in which the maximum correlation coefficients appear as latent roots and the scores $\{x_i\}$ and $\{y_j\}$ as latent vectors. The approach is reviewed by Lancaster (1969a, pp. 194–7), who criticizes the tests of significance which have been proposed.

(b) The use of scores is avoided altogether. For example, when R is binary and S is ordered, we can test the hypothesis that R and S are independent against the alternative that

$$\lambda_{11} \leqslant \lambda_{12} \leqslant \ldots \leqslant \lambda_{1,s-1},$$

where at least one of the inequalities is strict. A general account of statistical inference under order restrictions is given by Barlow, Bartholomew, Bremner and Brunk (1972). The same example can be formulated in Bayesian terms as the problem of finding the posterior probability that

$$\sum_{j=1}^{b} P_{1j} \geqslant \sum_{j=1}^{b} P_{2j} \quad (b = 1, 2, \ldots, s-1).$$

Methods of computing this probability are described by Weisberg (1972).

6.2 Logit regression

We now discuss an important special case of the first problem formulated in the previous section. Suppose that a binary response R is observed at each level of a quantitative factor $\mathfrak{F}$, and consider the model expressed by

$$\lambda_{1b} = \beta(x_b - x_s) \quad (b = 1, 2, \ldots, s-1).$$

Denote $\log(p_{1s}/p_{2s})$ by θ. Then

$$\log(p_{1j}/p_{2j}) = \theta + \beta(x_j - x_s) \quad (j = 1, 2, \ldots, s).$$

This is a linear regression model for logit p_{1j}.

We can make inferences about β either from the distribution of $\{N_{1j}\}$ conditional on the observed value of N_{10}, the values of $\{n_{0j}\}$ being specified, or from the unconditional distribution of $\{N_{1j}\}$. Consider first the conditional inferences. The distribution of $\{N_{1b}\}$ conditional on the observed value of N_{10} is $h\{n_{1b}; n_{i0}, n_{0j}, \lambda_{1b}\}$. When the expression for λ_{1b} is substituted in $h\{.\}$, we obtain a conditional likelihood function for β. The conditional maximum-likelihood estimator of β, say $\tilde{\beta}$, is determined by solving the equation

$$\Sigma_b N_{1b}(x_b - x_s) = E\{\Sigma_b N_{1b}(x_b - x_s); \tilde{\beta}\}.$$

On both sides,

$$\Sigma_b N_{1b}(x_b - x_s) = \Sigma_j N_{1j} x_j - n_{10} x_j.$$

The equation for $\tilde{\beta}$ is therefore also

$$\Sigma_j N_{1j} x_j = E(\Sigma_j N_{1j} x_j; \tilde{\beta}).$$

We similarly find that

$$\operatorname{var}(\tilde{\beta}) \sim \{\operatorname{var}(\Sigma_j N_{1j} x_j; \beta)\}^{-1}.$$

These results call for the mean and variance of $\Sigma_j N_{1j} x_j$, but no explicit expressions are available except when $\beta = 0$, in which case $\mathfrak{F}$ has no effect on R. We therefore turn to the unconditional distribution of $\{N_{ij}\}$. This is the product of s binomial distributions, the jth of which has probabilities

$$p_{1j} = \exp\{\theta + \beta(x_j - x_s)\}/[1 + \exp\{\theta + \beta(x_j - x_s)\}],$$

and

$$p_{2j} = 1/[1 + \exp\{\theta + \beta(x_j - x_s)\}].$$

Thus the kernel of the unconditional log likelihood function, $\log L$, is

$$\Sigma_j N_{1j}\{\theta + \beta(x_j - x_s)\} - \Sigma_j n_{0j} \log[1 + \exp\{\theta + \beta(x_j - x_s)\}].$$

The unconditional maximum-likelihood equations are

$$N_{10} = \Sigma_j n_{0j} \hat{p}_{1j} \quad \text{and} \quad \Sigma_j N_{1j} x_j = \Sigma_j n_{0j} \hat{p}_{1j} x_j,$$

in which

$$\hat{p}_{1j} = \exp\{\hat{\theta} + \hat{\beta}(x_j - x_s)\}/[1 + \exp\{\hat{\theta} + \hat{\beta}(x_j - x_s)\}].$$

There is no explicit solution of these equations. In such circumstances, the standard method of calculating the estimates is the Newton–Raphson procedure, which also generates the asymptotic dispersion matrix. This procedure is discussed in general terms in section 6.3, and its application to logit regression is described by Ashton (1972, ch. 4). The second-order derivatives of $\log L$ which are required are given by

$$\partial^2 \log L/\partial\theta^2 = \Sigma_j\, n_{0j} p_{1j} p_{2j},$$

$$-\partial^2 \log L/\partial\theta\,\partial\beta = \Sigma_j\, n_{0j} p_{1j} p_{2j}(x_j - x_s),$$

and
$$-\partial^2 \log L/\partial\beta^2 = \Sigma_j\, n_{0j} p_{1j} p_{2j}(x_j - x_s)^2.$$

Since they do not involve $\{N_{1j}\}$, their estimated values are the corresponding elements of the information matrix. A confidence region for (θ, β) can be obtained from a bivariate normal approximation to the distribution of $(\hat{\theta}, \hat{\beta})$.

Another method of solving the maximum-likelihood equations is the following. For an arbitrary value of β, define $m_{ij}(\beta)$ to satisfy the conditions

$$\Sigma_j\, m_{ij}(\beta) = n_{i0}, \quad \Sigma_i\, m_{ij}(\beta) = n_{0j},$$

and
$$\log \{m_{1b}(\beta) m_{2s}(\beta)/m_{1s}(\beta) m_{2b}(\beta)\} = \beta(x_b - x_s).$$

We have shown in sections 4.5 and 5.3 that m_{ij} is the unconditional maximum-likelihood estimate of the expected frequency in row i and column j. Thus,

$$m_{ij}(\hat{\beta}) = n_{0j} \hat{p}_{ij}.$$

Consider first the calculation of $\hat{\beta}$. An estimator of $\log (p_{1j}/p_{2j})$ is $\log (N_{1j}/N_{2j})$. We therefore begin by plotting the observed results on a graph with ordinate $\log (n_{1j}/n_{2j})$ amd abscissa x_j. If the model fits reasonably well, the plotted points conform approximately to a straight line. The slope of the line is a first approximation to $\hat{\beta}$. For values of β near this approximation, we use the iterative scaling procedure to calculate $m_{ij}(\beta)$. The appropriate system of cross-product ratios is defined by the following table:

$e^{\beta x_1}$	$e^{\beta x_2}$	...	$e^{\beta x_s}$
1	1	...	1

At each value of β, we also calculate

$$Y^2(\beta) = 2\, \Sigma_{ij}\, N_{ij} \log \{N_{ij}/m_{ij}(\beta)\}.$$

The maximum-likelihood estimator $\hat{\beta}$ is the value of β which minimizes $Y^2(\beta)$, and

$$\hat{\theta} = \log \{m_{1s}(\hat{\beta})/m_{2s}(\hat{\beta})\}.$$

Provided that the agreement between $\{N_{ij}\}$ and $\{m_{ij}(\hat{\beta})\}$ is satisfactory, we continue the analysis by constructing a confidence interval for β. Denote the minimum value of $Y^2(.)$ by $Y^2_{\min}$, and let k^2_ϵ be the upper 100ϵ percent point of χ^2_1. Then the set of values of β such that

$$Y^2(\beta) \leqslant Y^2_{\min} + k^2_\epsilon$$

is a confidence interval for β with confidence coefficient approximately $1 - \epsilon$.

Example 6.3 The data in Table 3.2 show how the mortality of *Tribolium castaneum* depends on time and the deposit of γ-benzene hexachloride. A binary response is obtained by pooling the last two categories of survival. The corresponding frequencies are given below, where the explanatory variable is $\log_{10}$ deposit.

$\log_{10}$ deposit	1·082	1·161	1·212	1·258	1·310	1·348
Dead	15	24	26	24	29	29
Alive	35	25	24	26	21	20

We obtain the following results. The minimum value of $Y^2(.)$ is 2·216 (4 d.f.). This value is attained when $\hat{\beta} = 3{\cdot}964$. The corresponding expected frequencies appear below.

17·617	20·905	23·833	26·111	28·662	29·871
32·382	28·095	26·167	23·889	21·338	19·129

A confidence interval for β with confidence coefficient approximately 95 percent is obtained by taking $k_\epsilon^2 = 3{\cdot}841$ and is given by (1·376, 6·619).

Finally, we mention a different estimation procedure with the advantage of freedom from iteration. Define the *logit chi-square* by

$$U = \Sigma_j (N_{1j}N_{2j}/n_{0j})\{\log (N_{1j}/N_{2j}) - \theta - \beta(x_j - x_s)\}^2.$$

The *minimum logit chi-square* estimators of θ and β are obtained by minimizing U, and can therefore be calculated from standard formulae for weighted linear regression. This procedure, due to Berkson (1944, 1955), is a special case of a general method discussed in the following section.

6.3 Linear models

Consider a set of mutually independent variates $N_1, N_2, \ldots, N_t$, such that N_i has a Poisson distribution with parameter μ_i ($i = 1, 2, \ldots, t$). The terminology is intended to include any of the experimental situations already discussed, and further generalizations of them. For example, the frequencies in a contingency table of order $r \times s$ can be written as a vector, and then distinguished by the values of a single suffix. Let $\phi(\mu)$ be a specified function of μ, which will usually be either $\log \mu$ or μ. Denote $\phi(\mu_i)$ by ϕ_i, and write $\boldsymbol{\phi}$ for the vector with components $\{\phi_i\}$. We assume that

$$\boldsymbol{\phi} = \mathbf{A}\boldsymbol{\theta},$$

where A is a specified matrix and $\boldsymbol{\theta}$ is a vector of u unknown parameters ($u \leqslant t$). This completes the specification of a linear model which provides a general basis for the analysis of categorical data.

We use maximum-likelihood procedures to estimate both $\boldsymbol{\theta}$ and the dispersion matrix of the vector estimator. The logarithm of the likelihood is a function of $\boldsymbol{\theta}$, expressed in terms of $\{\mu_i\}$ by

$$\log L = \text{const.} + \Sigma_i (N_i \log \mu_i - \mu_i).$$

A typical first-order derivative is

$$\frac{\partial \log L}{\partial \theta_j} = \Sigma_i \frac{\partial \log L}{\partial \mu_i} \frac{d\mu_i}{d\phi_i} \frac{\partial \phi_i}{\partial \theta_j} = \Sigma_i \left(\frac{N_i - \mu_i}{\mu_i}\right) \frac{a_{ij}}{\phi_i'},$$

where ϕ_i' stands for $\phi'(\mu_i)$. Thus the equations of estimation take the form

$$\Sigma_i (N_i - \mu_i) a_{ij} / \mu_i \phi_i' = 0.$$

When $\phi(\mu) = \log \mu$, the maximum-likelihood estimates are unique whenever they exist, and they exist if there are no zero frequencies (Birch, 1963; Haberman, 1970, ch. II). The equations of estimation cannot be solved explicitly. Many procedures have been developed which may be used to maximize a likelihood function (Chambers, 1973). In the present context, where the second-order derivatives of the log likelihood are calculated without difficulty, the Newton–Raphson method is a convenient iterative procedure of successive approximation. A typical second-order derivative is

$$\frac{\partial^2 \log L}{\partial \theta_j \partial \theta_k} = \Sigma_i \frac{a_{ij}}{\phi_i'} \frac{\partial}{\partial \theta_k} \left(\frac{N_i - \mu_i}{\mu_i}\right) + \Sigma_i \left(\frac{N_i - \mu_i}{\mu_i}\right) \frac{\partial}{\partial \theta_k} \left(\frac{a_{ij}}{\phi_i'}\right).$$

The expectation of the second term on the right is zero, whence

$$-E\,(\partial^2 \log L / \partial \theta_j\, \partial \theta_k) = \Sigma_i\, a_{ij} a_{ik} / \mu_i (\phi_i')^2.$$

Write w_i for $1/\mu_i(\phi_i')^2$. This coefficient is known as a *weight*. We therefore have

$$\partial \log L / \partial \theta_j = \Sigma_i\, w_i a_{ij} (N_i - \mu_i) \phi_i',$$

and
$$-E\,(\partial^2 \log L / \partial \theta_j\, \partial \theta_k) = \Sigma_i\, w_i a_{ij} a_{ik}.$$

The dispersion matrix of $\hat{\boldsymbol{\theta}}$ is estimated by the inverse of the matrix with elements $\{\Sigma_i\, \hat{w}_i a_{ij} a_{ik}\}$.

We now consider the Newton–Raphson method, in which each approximation to $\hat{\boldsymbol{\theta}}$ is succeeded by a closer approximation. Let θ_j^c be the approximation to $\hat{\theta}_j$ at stage c of the iterative process, and use a similar notation for the other quantities which change from stage to stage, namely w_i, μ_i, ϕ_i, and ϕ_i'. The approximation to $\hat{\theta}_j$ at stage $c + 1$ is

$$\theta_j^{c+1} = \theta_j^c + \delta \theta_j^c,$$

where the corrections $\{\delta\theta_j^c\}$ are such that

$$\Sigma_k\ \delta\theta_k^c\ \Sigma_i\ w_i^c a_{ij} a_{ik} = \Sigma_i\ w_i^c a_{ij}(n_i - \mu_i^c)\phi_i'^c.$$

However, the assumption that $\boldsymbol{\varphi} = \mathbf{A}\boldsymbol{\theta}$ implies that

$$\Sigma_k\ \theta_k^c\ \Sigma_i\ w_i^c a_{ij} a_{ik} = \Sigma_i\ w_i^c a_{ij}\ \phi_i^c.$$

Therefore the components $\{\theta_j^{c+1}\}$ satisfy the system of equations

$$\Sigma_k\ \theta_k^{c+1}\ \Sigma_i\ w_i^c\ a_{ij} a_{ik} = \Sigma_i\ w_i^c\ a_{ij}\{\phi_i^c + (n_i - \mu_i^c)\phi_i'^c\}.$$

A first approximation to $\hat{\boldsymbol{\theta}}$ can sometimes be obtained by graphical methods. In general, the procedure is started by taking $\mu_i^1 = n_i$, which provides initial values for ϕ_i, ϕ_i', and w_i. We then continue the iterations until successive approximations agree within prescribed limits. The same method is used in probit analysis (Finney, 1971, sect. 4.2), and a generalized version, which can be applied to any distribution in the exponential family, is described by Nelder and Wedderburn (1972). A disadvantage of the traditional Newton–Raphson algorithm is failure to converge with some starting values. The algorithm can be modified to remove this disadvantage (Haberman, 1970, ch. III).

Denote by $\boldsymbol{\theta}^*$ the approximation to $\hat{\boldsymbol{\theta}}$ obtained by taking $\mu_i^1 = n_i$. The components of $\boldsymbol{\theta}^*$ satisfy the system of equations

$$\Sigma_k\ \theta_k^*\ \Sigma_i\ a_{ij} a_{ik}/n_i\{\phi'(n_i)\}^2 = \Sigma_i\ a_{ij}\phi(n_i)/n_i\{\phi'(n_i)\}^2.$$

Exactly the same system is obtained if we suppose (inaccurately) that

$$E\{\boldsymbol{\varphi}(N)\} = \mathbf{A}\boldsymbol{\theta},$$

estimate var $\{\phi(N_i)\}$ by the asymptotic formula $N_i\{\phi'(N_i)\}^2$, and minimize

$$W = \Sigma_i\ \{\phi(N_i) - \Sigma_j\ a_{ij}\theta_j\}^2/N_i\{\phi'(N_i)\}^2$$

with respect to the parameters $\{\theta_j\}$. This is an application of the principle of least squares with empirical weights, and standard techniques can be used to make inferences about $\boldsymbol{\theta}$. In particular, the equations of estimation are linear, the dispersion matrix of the estimators is the inverse of the matrix of the equations of estimation, and the test statistic for a linear hypothesis H is the difference between (i) the relative minimum of W under H, and (ii) the absolute minimum of W. Similar procedures have been advocated by Berkson since 1944 for the case when $\phi(\mu) = \log \mu$. A recent statement of his approach is contained in Berkson (1972), and the same ideas are formulated in general terms by Grizzle, Starmer and Koch (1969). When $\phi(\mu) = \mu$,

$$W = \Sigma_i\ (N_i - \Sigma_j\ a_{ij}\theta_j)^2/N_i,$$

which differs from Pearson's statistic in that the denominator of each term is an observed frequency instead of an expected frequency. For this reason, the procedure of minimizing W is known as the *minimum modified chi-square* method.

The arguments given to support the use of minimum modified chi-square estimators are as follows:

(1) The calculations are much less than those called for by maximum likelihood. On the other hand, while there will always be times and circumstances when ease of calculation is important, the increasing availability of computers has greatly diminished the force that this argument had in the past.

(2) The minimum modified chi-square estimator $\boldsymbol{\theta}^*$ is best asymptotically normal, in the sense that $\boldsymbol{\theta}^*$ and the maximum-likelihood estimator $\hat{\boldsymbol{\theta}}$ have the same asymptotic properties of consistency, efficiency, and normality (Neyman, 1949). On the other hand, the second-order asymptotic efficiency of $\hat{\boldsymbol{\theta}}$ exceeds that of $\boldsymbol{\theta}^*$ and is the maximum possible (Rao, 1961, Table 1). However, the concept of second-order efficiency is somewhat remote from practical realities, where $\boldsymbol{\theta}^*$ and $\hat{\boldsymbol{\theta}}$ are really indistinguishable.

6.4 Log-linear models

We now discuss the linear model obtained by taking

$$\phi(\mu) = \log \mu$$

and given by

$$\log \mu_i = \Sigma_j \, a_{ij}\theta_j \quad (i = 1, 2, \ldots, t).$$

This defines a log-linear model more general in scope than the type encountered in section 3.4, where the coefficients $\{a_{ij}\}$ are restricted to the values 0 and 1. The log likelihood becomes

$$\log L = \text{const.} + \Sigma_{ij} \, N_i a_{ij}\theta_j - \Sigma_i \exp(\Sigma_j \, a_{ij}\theta_j).$$

Denote $\Sigma_i \, N_i a_{ij}$ by S_j, so that the kernel of the log likelihood is

$$\Sigma_j \, S_j\theta_j - \Sigma_i \exp(\Sigma_j \, a_{ij}\theta_j).$$

Thus the statistics $\{S_j\}$ are minimal sufficient for the parameters $\{\theta_j\}$. The equations of estimation take the form

$$S_j = E(S_j; \hat{\theta}) \quad (j = 1, 2, \ldots, u).$$

Denote the dispersion matrices of $\hat{\boldsymbol{\theta}}$ and $(S_1, S_2, \ldots, S_u)$ by $\mathfrak{D}(\hat{\boldsymbol{\theta}})$ and $\mathfrak{D}(\mathbf{S})$ respectively. Then

$$\mathfrak{D}(\hat{\boldsymbol{\theta}}) \sim \{\mathfrak{D}(\mathbf{S})\}^{-1}.$$

When the coefficients $\{a_{ij}\}$ are either 0 or 1, the equations of estimation assert that certain marginal totals are the same for both the observed and expected frequencies. Given a set of assigned values for the cross-product ratios, the equations can be solved for the $\{\mu_i\}$. The iterative scaling procedure (sect. 3.4) is then a more convenient method of calculation than the Newton–Raphson procedure because no matrix inversion is involved. Convergence is first-order for iterative scaling as compared with second-order for Newton–Raphson, but a few iterations usually suffice, and many of the calculations are within the range of a desk-top calculator. An algorithm in ANSI Standard Fortran is given by Haberman (1972).

Example 6.4 The *rs* variates $\{N_{ij}\}$ are such that

$$\log \mu_{rs} = \kappa, \quad \log \mu_{as} = \kappa + \alpha_a, \quad \log \mu_{rb} = \kappa + \beta_b,$$

and

$$\log \mu_{ab} = \kappa + \alpha_a + \beta_b,$$

where $a = 1, 2, \ldots, r-1$ and $b = 1, 2, \ldots, s-1$.
Sufficient statistics are the total frequency N, together with $\{N_{ao}\}$ and $\{N_{ob}\}$. The equations of estimation are

$$\hat{\mu} = N, \quad \hat{\mu}_{ao} = N_{ao}, \quad \text{and} \quad \hat{\mu}_{ob} = N_{ob},$$

where $\hat{\mu}$ denotes $\Sigma_{ij}\, \hat{\mu}_{ij}$. A set of conditions implied by the model is

$$\log (\mu_{ab}\mu_{rs}/\mu_{as}\mu_{rb}) = 0 \quad \text{for all } a \text{ and } b.$$

There is an explicit solution for the maximum-likelihood estimators, which are given by

$$\hat{\mu}_{ij} = N_{io}N_{oj}/N.$$

In general, the parameters $\{\theta_j\}$ are regression coefficients, and the quantities $\{a_{ij}\}$ are values of explanatory variables. The Newton–Raphson method is usually the best procedure for solving the equations of estimation. However, a combination of iterative scaling and search procedures can be used when the number of regression parameters is small, and estimates of parameters and their dispersion matrix can be obtained from the table of expected frequencies. Other procedures for maximum-likelihood estimation from the log-linear model have been proposed (Darroch and Ratcliff, 1972; Gokhale, 1972).

The property of joint sufficiency implies that the distribution of S_1 conditional on the observed values of $S_2, S_3, \ldots, S_u$ depends only on θ_1, the other parameters being eliminated. This distribution is given by

$$\{\exp (s_1\theta_1)/\Pi_i\, n_i!\}/\{\Sigma \exp (s_1\theta_1)/\Pi_i\, n_i!\},$$

where the sum in the denominator is taken over all non-negative integers $\{n_i\}$ such that

$$\Sigma_i\, n_i a_{ij} = s_j \quad (j = 2, 3, \ldots, u).$$

We find as before that the conditional maximum-likelihood estimator of θ_1 satisfies the equation

$$S_1 = E\,(S_1 \,|\, S_2 = s_2, \ldots, S_u = s_u; \widetilde{\theta}_1),$$

and similarly for $\theta_2, \theta_3, \ldots, \theta_u$.

Example 6.5 The variates $N_{11}, N_{12}, N_{21}, N_{22}$ are such that

$$\log \mu_{22} = \theta_1, \quad \log \mu_{12} = \theta_1 + \theta_2, \quad \log \mu_{21} = \theta_1 + \theta_3,$$

and $$\log \mu_{11} = \theta_1 + \theta_2 + \theta_3 + \theta_4.$$

Here $S_1 = N$, $S_2 = N_{10}$, $S_3 = N_{01}$, and $S_4 = N_{11}$.

Inferences about θ_4 are made from the distribution $h(n_{11}; n_{i0}, n_{0j}, \theta_4)$, where

$$\theta_4 = \log\,(\mu_{11}\mu_{22}/\mu_{12}\mu_{21}).$$

When the frequencies $\{N_i\}$ result from the simultaneous use of different methods of classification, at least one of which is binary, the log-linear model can be reformulated as follows. A change of notation is convenient, and we replace $\{N_i\}$ by $\{N_{fg}\}$, where $f = 1, 2$ and $g = 1, 2, \ldots, v$. The frequencies can be arranged in a contingency table of order $2 \times v$, the rows of which define a binary classification. We assume that

$$\log \mu_{1g} = \beta_g + \Sigma_j\, b_{gj}\theta_j \quad \text{and} \quad \log \mu_{2g} = \beta_g + \Sigma_j\, c_{gj}\theta_j.$$

Here $\beta_1, \beta_2, \ldots, \beta_v$ are parameters attached to the columns. They correspond to block effects in the analysis of variance, and we require to eliminate them in order to make inferences about $\boldsymbol{\theta}$. We therefore consider the joint distribution of $\{N_{fg}\}$ conditional on the observed values of the column totals $\{N_{0g}\}$. This is a product of v binomial distributions, typified by the distribution of N_{1g} conditional on $N_{0g} = n_{0g}$, which has index n_{0g} and probability p_g, where

$$p_g = \mu_{1g}/(\mu_{1g} + \mu_{2g}).$$

Write a_{gj} for $b_{gj} - c_{gj}$, and we obtain

$$\operatorname{logit} p_g = \Sigma_j\, a_{gj}\theta_j.$$

This is the linear *logistic* model, which is discussed at length by Cox (1970a). The model successfully unifies a wide range of statistical techniques concerned with the analysis of binary data, but has the following disadvantages:

(1) The binary classification to which the probabilities $\{p_g\}$ refer is placed in a special position. Thus, when several binary classifications are present, there is a choice of logistic models, and the question arises as to which of

them gives the best representation of the data.
(2) When three or more methods of classification are used, the expected frequencies for various relevant hypotheses can be calculated by the iterative scaling procedure, whereas any derived logistic model calls for the Newton–Raphson procedure. A comparative study of different ways of fitting log-linear models to multivariate categorical data with at least one binary response is made by Bishop (1969).

7 THREE-WAY CLASSIFICATIONS

7.1 Three-dimensional contingency tables

The contingency tables discussed in previous chapters are *two-dimensional* because the frequencies observed are classified by two methods. When another method of classification is introduced, the frequencies define a *three-dimensional* contingency table. The description is purely formal, and various ways of presenting the data are illustrated in the examples below.

Suppose that $R, S, T, \ldots$ are responses and $\mathfrak{F}, \mathfrak{G}, \ldots$ are factors. There are two basic types of two-dimensional contingency table, depending on whether a univariate response is observed at each level of a factor, or whether a bivariate response is observed. The corresponding experimental situations may be denoted by $R\mathfrak{F}$ and RS respectively. In the same way, there are three basic types of three-dimensional contingency table:

(a) $R\mathfrak{F}\mathfrak{G}$, where a univariate response is observed for each treatment combination of two factors;

(b) $RS\mathfrak{F}$, where a bivariate response is observed at each level of a factor;

(c) RST, where a trivariate response is observed.

Example 7.1 The data in Table 7.1 are quoted from Hoblyn and Palmer (1934). They refer to an experiment designed to investigate the propagation

TABLE 7.1 *Survival of root cuttings in relation to length of cutting and time of planting*

Length of cutting	Time of planting			
	At once		In spring	
	Alive	Dead	Alive	Dead
Long	156	84	84	156
Short	107	133	31	209

of plum root-stocks from root cuttings. Plants are classified as alive or dead, and this defines the response. There are two factors, which each occur at two levels, as follows:

Length of cutting: long/short.
Time of planting : at once/in spring.

The number of cuttings is 240 for each treatment combination. Inspection of the table suggests that long cuttings survive better than short, and planting at once is preferable to planting in spring. The problem is to quantify these statements, and find whether or not the interaction of the treatments affects the response.

Example 7.2 The data in Table 7.2 are quoted by Cochran (1954). Mothers of children who had been referred by their teachers as presenting behavioural

TABLE 7.2 *Losses of mothers in relation to behavioural problems and birth-order of children*

Numbers of mothers with	Birth order 2		Birth order 3–4		Birth order 5+	
	Problems	Controls	Problems	Controls	Problems	Controls
Losses	20	10	26	16	27	14
None	82	54	41	30	22	23
Percent loss	19·6	15·6	38·8	34·8	55·1	37·8

problems are compared with mothers of children who had not been so referred. The two groups are described respectively as problems and controls. A record was made for each mother of whether or not she had suffered any infant losses (e.g. stillbirths) prior to the birth of the child in the study. The classification losses/none defines a binary response. Since the loss rate increases with the birth-order of the child, data are given for three birth-order classes. Group and birth-order are treated as factors.

Example 7.3 The data in Table 7.3 are quoted by Stuart (1953). They are based on case-records of the eye-testing of men employees in Royal

TABLE 7.3 *Unaided distance vision of 3242 men aged 30–39*

Grade of right eye	Grade of left eye: Highest	Second	Third	Lowest	Total
Highest	821	112	85	35	1053
Second	116	494	145	27	782
Third	72	151	583	87	893
Lowest	43	34	106	331	514
Total	1052	791	919	480	3242

Ordnance factories in 1943–6. The corresponding data for women are given in Table 3.5. There are two responses, defined by vision in the right and left eyes. A question which arises is whether the association between responses is the same for men and women.

Example 7.4 The data in Table 7.4 are given by Ashford and Sowden (1970). They refer to coalminers who were smokers without radiological

TABLE 7.4 *Coalminers classified by age, breathlessness, and wheeze*

Age-group in years	Breathlessness		No breathlessness		Total
	Wheeze	No wheeze	Wheeze	No wheeze	
20–24	9	7	95	1841	1952
25–29	23	9	105	1654	1791
30–34	54	19	177	1863	2113
35–39	121	48	257	2357	2783
40–44	169	54	273	1778	2274
45–49	269	88	324	1712	2393
50–54	404	117	245	1324	2090
55–59	406	152	225	967	1750
60–64	372	106	132	526	1136
Total	1827	600	1833	14 022	18 282

pneumoconiosis and between the ages of 20 and 64 years (inclusive) at the time of examination. A short questionnaire about respiratory symptoms was given to each man, and the replies obtained enabled him to be classified with respect to the presence or absence of specific symptoms. The symptoms considered here are breathlessness and wheeze. A subdivision of the miners into five-year age-groups was also made. We therefore have two binary responses, and a third response determined quantitatively. Inspection of the data shows that the prevalence of each symptom increases with increasing age. One object of the analysis is to place this observation on a quantitative basis. The question also arises whether the association between breathlessness and wheeze varies systematically with age and, if so, in what way.

7.2 Association of three responses

Suppose that a population is classified by each of three methods, which contain r, s, and t categories respectively. The corresponding responses R, S, and T have a joint distribution given by

$$p(R=i, S=j, T=k) \;=\; p_{ijk}$$

$$(i = 1,2,\ldots,r; j = 1,2,\ldots,s; k = 1,2,\ldots,t).$$

We discussed the association between two responses in section 3.3 and were led to an analysis based on a transformation of the probabilities. The association of three responses can be analysed by similar methods.

We assume that $p_{ijk} > 0$ for all i, j, and k. Define parameters as follows.

$$\kappa = \log p_{rst}, \quad \theta_a = \log (p_{ast}/p_{rst}), \quad \phi_b = \log (p_{rbt}/p_{rst}),$$

$$\lambda_{ab} = \log (p_{abt}p_{rst}/p_{ast}p_{rbt}), \quad \psi_c = \log (p_{rsc}/p_{rst}),$$

$$\mu_{ac} = \log (p_{asc}p_{rst}/p_{ast}p_{rsc}), \quad \nu_{bc} = \log (p_{rbc}p_{rst}/p_{rbt}p_{rsc}),$$

and $$\xi_{abc} = \log (p_{abc}p_{ast}p_{rbt}p_{rsc}/p_{abt}p_{asc}p_{rbc}p_{rst}).$$

Here $a = 1, 2, \ldots, r-1$, $b = 1, 2, \ldots, s-1$, and $c = 1, 2, \ldots, t-1$.

The probabilities are expressed in terms of the parameters by the system of equations given below

$$\log p_{rst} = \kappa, \quad \log p_{ast} = \kappa + \theta_a, \quad \log p_{rbt} = \kappa + \phi_b,$$

$$\log p_{abt} = \kappa + \theta_a + \phi_b + \lambda_{ab}, \quad \log p_{rsc} = \kappa + \psi_c,$$

$$\log p_{asc} = \kappa + \theta_a + \psi_c + \mu_{ac}, \quad \log p_{rbc} = \kappa + \phi_b + \psi_c + \nu_{bc},$$

and $$\log p_{abc} = \kappa + \theta_a + \phi_b + \psi_c + \lambda_{ab} + \mu_{ac} + \nu_{bc} + \xi_{abc}.$$

This system constitutes a log-linear model of the form developed by Mantel (1966). An earlier version by Birch (1963) is analogous to the linear model for a three-way layout in the analysis of variance, and is defined as follows:

$$\log p_{ijk} = \mu + \alpha_i + \beta_j + \gamma_k + \delta_{ij} + \eta_{ik} + \zeta_{jk} + \omega_{ijk},$$

where $\Sigma_i\, \alpha_i$, $\Sigma_i\, \delta_{ij}$, $\Sigma_i\, \eta_{ik}$, $\Sigma_i\, \omega_{ijk}$ are all zero, and similarly for the suffices j and k. We shall continue to use the later version, which has the advantage that the parameters are expressed more simply in terms of the probabilities.

Consider the hypothesis

$$H_\xi : \xi_{abc} = 0 \quad \text{for all } a, b \text{ and } c.$$

When H_ξ holds,

$$\log (p_{ibc}p_{ist}/p_{ibt}p_{isc}) = \nu_{bc} \quad \text{for all } i.$$

Thus the system of cross-product ratios for the contingency table of order $s \times t$ obtained when $R = i$ is the same for all values of i. A similar argument shows that the system of cross-product ratios is the same for each category of S, and for each category of T. We have explained that the association between two responses can reasonably be measured by the corresponding set of cross-product ratios. Hence the association between R and S is the same for each category of T and similarly when the responses R, S, and T are cyclicly permuted. The term "interaction" is a synonym for "association". We therefore describe H_ξ as the hypothesis that the *second-order* or *three-response interaction* $R \times S \times T$ is zero (Bartlett, 1935; Roy and Kastenbaum, 1956). Consequently the parameters $\{\xi_{abc}\}$ together give a general definition of $R \times S \times T$. This conclusion receives further support from arguments based on symmetry (Altham, 1970b).

The definition of zero three-response interaction by the hypothesis H_ξ has been criticized on several occasions by Lancaster (1967; 1969a, p. 274; 1969b). His objections can be expressed as follows. The probabilities $\{p_{ijk}\}$ define a three-dimensional contingency table in which p_{ijk} occupies the cell formed by the ith row, jth column, and kth layer. Suppose that the three-response interaction is zero. Now consider a new table which is formed by collapsing over some of the layers, leaving the remainder unchanged. Numerical examples show that the new table can display a three-response interaction. The criticism is: with less information about the variates, we can detect an interaction not present before pooling.

An answer to this criticism is as follows. The relationship between the probabilities in a layer is expressed by row and column effects as well as by any interactions that may be present. Pooling is unjustified if this relationship varies between layers, as it clearly does in the examples which Lancaster gives. On the other hand, pooling is justified if the layers are completely homogeneous. Similar questions arise in connection with crude death-rates (Armitage, 1971, sect. 12.6).

Consider a table which has been collapsed over all the layers. Denote by ϵ_{ab} the function of the marginal probabilities corresponding to λ_{ab}, so that

$$\epsilon_{ab} = \log (p_{ab0}p_{rs0}/p_{as0}p_{rb0}).$$

When H_ξ is true, $\lambda_{ab} = \epsilon_{ab}$ for all a and b only if the table is *perfect* (Darroch, 1962), where a perfect table is one which satisfies

$$p_{ijk} = p_{0jk}p_{i0k}p_{ij0}/p_{i00}p_{0j0}p_{00k} \quad \text{for all } i, j, \text{ and } k.$$

Thus, when the three-response interaction is zero, we cannot give a clear meaning to two-response interactions defined in terms of marginal probabilities, and must therefore exercise care in selecting the sequence of hypotheses to be tested. Further aspects of pooling are discussed by Bishop (1971).

The following procedure (Goodman, 1970) provides a nested sequence of hypotheses and an interpretation of the parameters other than $\{\xi_{abc}\}$. Suppose that H_ξ holds, and consider the hypothesis

$$H_\nu : \nu_{bc} = 0 \quad \text{for all } b \text{ and } c.$$

H_ξ and H_ν together imply that

$$p_{ibc}p_{ist} = p_{ibt}p_{isc} \quad \text{for all } i, b, \text{ and } c.$$

Denote $p(S = j, T = k | R = i)$ by q_{jk}. By definition, $q_{jk} = p_{ijk}/p_{i00}$, and therefore

$$q_{bc}q_{st} = q_{bt}q_{sc} \quad \text{for all } b \text{ and } c.$$

This expresses the hypothesis of independent responses for a two-dimensional table with probabilities $\{q_{jk}\}$. Hence

$$q_{jk} = q_{jo}q_{ok} \quad \text{for all } j \text{ and } k,$$

or equivalently,

$$p_{ijk} = p_{ijo}p_{iok}/p_{ioo} \quad \text{for all } i, j \text{ and } k.$$

The conjunction of H_ξ and H_ν therefore admits the following interpretation: for each value of i, S and T are independent, conditional on $R = i$.

Suppose next that H_ξ and H_ν both hold. Consider the hypothesis

$$H_\mu : \mu_{ac} = 0 \quad \text{for all } a \text{ and } c.$$

A repetition of the argument used when combining H_ξ and H_ν shows that H_ξ and H_μ together imply that

$$p_{ijk} = p_{ijo}p_{ojk}/p_{ojo} \quad \text{for all } i, j \text{ and } k.$$

When H_ξ, H_ν and H_μ all hold, we obtain

$$p_{iok}p_{ojo} = p_{ojk}p_{ioo} \quad \text{for all } i, j \text{ and } k.$$

Summation over i leads to

$$p_{ojk} = p_{ojo}p_{ook} \quad \text{for all } j \text{ and } k.$$

Thus H_ξ, H_ν and H_μ together imply that

$$p_{ijk} = p_{ijo}p_{ook} \quad \text{for all } i, j \text{ and } k.$$

The addition of H_μ to the conjunction of H_ξ and H_ν is therefore equivalent to the hypothesis that

$$p_{iok} = p_{ioo}p_{ook} \quad \text{for all } i \text{ and } k.$$

This is the condition that R and T are independent.

The final stage of the sequence now follows. Suppose that H_ξ, H_ν and H_μ all hold. Consider the hypothesis

$$H_\lambda : \lambda_{ab} = 0 \quad \text{for all } a \text{ and } b.$$

A repetition of the arguments in the previous stage shows that H_ξ, H_ν and H_λ together imply that

$$p_{ijk} = p_{iok}p_{ojo} \quad \text{for all } i, j \text{ and } k.$$

When H_ξ, H_ν, H_μ and H_λ all hold, we obtain

$$p_{ijk} = p_{ioo}p_{ojo}p_{ook} \quad \text{for all } i, j \text{ and } k.$$

The addition of H_λ to the conjunction of H_ξ, H_ν and H_μ is therefore equivalent to the hypothesis that

$$p_{ij0} = p_{i00}p_{0j0} \quad \text{for all } i \text{ and } j.$$

This is the condition that R and S are independent.

7.3 Distribution theory

We consider a population classified by each of three methods, which define responses R, S, and T. A sample of size n is taken and n_{ijk} is the frequency of the event $(R = i, S = j, T = k)$. The frequencies $\{n_{ijk}\}$ form a contingency table of order $r \times s \times t$, in which n_{ijk} occupies the cell formed by the ith row, jth column, and kth layer. Denote by N_{ijk} the variate corresponding to n_{ijk}. Then the distribution of $\{N_{ijk}\}$ is multinomial with index n and probabilities $\{p_{ijk}\}$. An alternative derivation of the distribution is as follows. Consider rst mutually independent variates $\{N_{ijk}\}$ such that N_{ijk} has a Poisson distribution with parameter μ_{ijk}. Denote $\Sigma_{ijk} N_{ijk}$ by N and $\Sigma_{ijk} \mu_{ijk}$ by μ. A sample of size n determines the condition $N = n$. The distribution of $\{N_{ijk}\}$ conditional on $N = n$ is the stated multinomial, with $p_{ijk} = \mu_{ijk}/\mu$.

When the multinomial distribution is expressed in terms of the parameters defined in the previous section, we obtain

$$n!\,\Pi_{ijk}\, p_{ijk}^{n_{ijk}}/n_{ijk}! = (n!/\Pi_{ijk}\, n_{ijk}!) \exp(n\kappa + \Sigma_a\, n_{a00}\theta_a + \Sigma_b\, n_{0b0}\phi_b$$

$$+ \Sigma_c\, n_{00c}\psi_c + \Sigma_{ab}\, n_{ab0}\lambda_{ab} + \Sigma_{ac}\, n_{a0c}\mu_{ac} + \Sigma_{bc}\, n_{0bc}\nu_{bc} + \Sigma_{abc}\, n_{abc}\xi_{abc}).$$

As before, the convention is that a suffix over which summation takes place is replaced by zero. We can verify that the two expressions for the distribution are identical by checking that the powers of p_{abc}, p_{abt}, p_{ast} and p_{rst} agree, because symmetry then ensures that the remaining powers will also agree. For example, the coefficient of $\log p_{ast}$ in the exponential term is

$$n_{a00} - \Sigma_b\, n_{ab0} - \Sigma_c\, n_{a0c} + \Sigma_{bc}\, n_{abc} = n_{ast}.$$

Corresponding to the sequence of hypotheses H_ξ, H_ν, H_μ and H_λ, there is a sequence of conditional distributions which we now derive. First, consider the hypothesis H_ξ. Inferences about $\{\xi_{abc}\}$ are based on the distribution of $\{N_{abc}\}$ conditional on the observed values of $\{N_{a00}\}$, $\{N_{0b0}\}$, $\{N_{00c}\}$, $\{N_{ab0}\}$, $\{N_{a0c}\}$ and $\{N_{0bc}\}$. Since the total frequency is fixed, the conditions are equivalent to the observed values of $\{N_{ij0}\}$, $\{N_{i0k}\}$, and $\{N_{0jk}\}$. The addition of $\{N_{abc} = n_{abc}\}$ implies that $\{N_{ijk} = n_{ijk}\}$. Thus the conditional distribution required is

$$\{\exp(\Sigma_{abc}\, n_{abc}\xi_{abc})/\Pi_{ijk}\, n_{ijk}!\}/\{\Sigma_{n_{abc}} \exp(\Sigma_{abc}\, n_{abc}\xi_{abc})/\Pi_{ijk}\, n_{ijk}!\}.$$

Secondly, suppose that H_ξ holds and consider H_ν. Inferences about $\{\nu_{bc}\}$ are based on the distribution of $\{N_{0bc}\}$ conditional on the observed values of $\{N_{a0c}\}$, $\{N_{ab0}\}$, $\{N_{a00}\}$, $\{N_{0b0}\}$ and $\{N_{00c}\}$. The conditions are equivalent to fixed values for $\{N_{ijo}\}$ and $\{N_{iok}\}$. For each value of i, the distribution of $\{N_{ijk}\}$ conditional on the observed values of $\{N_{ijo}\}$ and $\{N_{iok}\}$ is $h\{n_{ibc}; n_{ijo}, n_{iok}, \nu_{bc}\}$. The conditional distribution of $\{N_{obc}\}$ is the convolution of the conditional distributions of $\{N_{ibc}\}$ for $i = 1,2, \ldots, r$ (Birch, 1965)

Thirdly, suppose that H_ξ and H_ν both hold, and consider H_μ. Inferences about $\{\mu_{ac}\}$ are based on the distribution of $\{N_{aoc}\}$ conditional on the observed values of $\{N_{abo}\}$, $\{N_{aoo}\}$, $\{N_{obo}\}$, and $\{N_{ooc}\}$. The conditions are equivalent to the observed values of $\{N_{ijo}\}$ and $\{N_{ook}\}$. When the values of $\{N_{aoc}\}$ are also given, only the frequencies $\{n_{abc}\}$ and $\{n_{obc}\}$ remain variable. For each i,

$$\Sigma_{n_{ibc}} (1/\Pi_{jk}\, n_{ijk}!) = n_{ioo}!/\Pi_{jk}\, n_{ijo}!\, n_{iok}!$$

Thus the conditional distribution required is $h\{n_{aoc}; n_{ioo}, n_{ook}, \mu_{ac}\}$.

Finally, suppose that H_ξ, H_ν and H_μ all hold, and consider H_λ. Inferences about $\{\lambda_{ab}\}$ are based on the distribution of $\{N_{abo}\}$ conditional on the observed values of $\{N_{ioo}\}$ and $\{N_{ojo}\}$, namely $h\{n_{abo}; n_{ioo}, n_{ojo}, \lambda_{ab}\}$.

The conditional distributions which arise in the analysis of a three-dimensional contingency table include several extended hypergeometric distributions, which can be used to make inferences by methods already discussed. There also arises a generalized form of the extended hypergeometric distribution, namely

$$g\{n_{abc}; n_{ijo}, n_{iok}, n_{ojk}, \xi_{abc}\}$$
$$= \{\exp(\Sigma_{abc}\, n_{abc}\xi_{abc}/\Pi_{ijk}\, n_{ijk}!\}/\{\Sigma_{n_{abc}} \exp(\Sigma_{abc}\, n_{abc}\xi_{abc})/\Pi_{ijk}\, n_{ijk}!\}.$$

The exact use of $g\{.\}$ presents considerable problems of enumeration except in special cases, e.g. when $r = s = t = 2$ or when n is small. Recourse to asymptotic methods is nearly always essential. A normal multivariate approximation to the distribution of $\{N_{abc}\}$ is usually employed. The corresponding limit theorem is based on the property of three-dimensional contingency tables which now follows.

Suppose that the distribution of a trivariate response (R, S, T) is defined by

$$p(R = i, S = j, T = k) = p_{ijk}.$$

Then the probabilities $\{p_{ijk}\}$ are uniquely determined by a combination which consists of the two-dimensional marginal probabilities $\{p_{ijo}\}$, $\{p_{iok}\}$ and $\{p_{ojk}\}$, together with the parameters $\{\xi_{abc}\}$. Furthermore, the $\{p_{ijk}\}$ can be obtained by iterative scaling from any three-dimensional table with the specified values of $\{\xi_{abc}\}$, e.g. the table with elements $\{q_{ijk}\}$, where

$$q_{abc} = \exp(\xi_{abc}) \quad \text{and} \quad q_{ijk} = 1 \text{ otherwise.}$$

These results are proved by the same techniques which can be used to show that a two-dimensional contingency table is uniquely determined by a combination of the marginal totals and cross-product ratios (Ireland and Kullback, 1968).

The distribution $g\{.\}$ converges to normality under the following conditions. Let $\{\pi_{ij0}\}$, $\{\pi_{i0k}\}$, and $\{\pi_{0jk}\}$ be positive quantities which form a consistent set of two-dimensional marginal probabilities. Thus

$$\Sigma_j \pi_{ij0} = \Sigma_k \pi_{i0k}, \quad \Sigma_i \pi_{ij0} = \Sigma_k \pi_{0jk}, \quad \Sigma_i \pi_{i0k} = \Sigma_j \pi_{0jk},$$

and

$$\Sigma_{ij} \pi_{ij0} = \Sigma_{ik} \pi_{i0k} = \Sigma_{jk} \pi_{0jk} = 1.$$

For specified values of $\{\xi_{abc}\}$, define $m_{ijk}(\boldsymbol{\xi})$ to satisfy

$$\Sigma_i m_{ijk}(\boldsymbol{\xi}) = n_{0jk}, \quad \Sigma_j m_{ijk}(\boldsymbol{\xi}) = n_{i0k}, \quad \Sigma_k m_{ijk}(\boldsymbol{\xi}) = n_{ij0},$$

and

$$\log \left\{ \frac{m_{abc}(\boldsymbol{\xi})\, m_{ast}(\boldsymbol{\xi})\, m_{rbt}(\boldsymbol{\xi})\, m_{rsc}(\boldsymbol{\xi})}{m_{abt}(\boldsymbol{\xi})\, m_{asc}(\boldsymbol{\xi})\, m_{rbc}(\boldsymbol{\xi})\, m_{rst}(\boldsymbol{\xi})} \right\} = \xi_{abc}.$$

Consider the distribution $g\{.\}$ in the limit when $n_{ij0} \to \infty$, $n_{i0k} \to \infty$, and $n_{0jk} \to \infty$ in such a way that $n_{ij0}/n \to \pi_{ij0}$, $n_{i0k}/n \to \pi_{i0k}$, and $n_{0jk}/n \to \pi_{0jk}$ for all i, j, and k. Then

$$g\{.\} \sim q\{m_{ijk}(\boldsymbol{\xi})\} \exp\left[-\tfrac{1}{2} \Sigma_{ijk} \{n_{ijk} - m_{ijk}(\boldsymbol{\xi})\}^2 / m_{ijk}(\boldsymbol{\xi})\right].$$

The expression on the right is the density function of a singular normal multivariate distribution for the $(r-1)(s-1)(t-1)$ variates $\{N_{abc}\}$ in rst dimensions. Singularity arises because

$$\Sigma_i \{n_{ijk} - m_{ijk}(\boldsymbol{\xi})\} = 0 \quad \text{for all } j \text{ and } k,$$

and similarly for summations over j and k.

The method of proof is a further generalization of the corresponding methods for 2×2 tables in section 4.2 and $r \times s$ tables in section 5.1. At the stage where the contributions to the expansion of the log probability-ratio are collected, we use the property of $\{m_{ijk}(\boldsymbol{\xi})\}$ that

$$\Sigma_{ijk} \{n_{ijk} - m_{ijk}(\boldsymbol{\xi})\} \log m_{ijk}(\boldsymbol{\xi}) - \Sigma_{abc} \{n_{abc} - m_{abc}(\boldsymbol{\xi})\} \xi_{abc} \equiv 0.$$

This identity is verified by checking that the coefficients of $\log m_{abc}(\boldsymbol{\xi})$, $\log m_{abt}(\boldsymbol{\xi})$, $\log m_{ast}(\boldsymbol{\xi})$, and $\log m_{rst}(\boldsymbol{\xi})$ are all zero.

The asymptotic dispersion matrix of $\{N_{ijk}\}$ conditional on the observed two-dimensional totals is obtained by a straightforward extension of the method described for $r \times s$ tables in section 5.1.

7.4 Experiments involving factors

We now discuss two other experimental situations which give rise to a contingency table of order $r \times s \times t$. First, suppose that a bivariate response (S, T) is observed at each level of a factor $\mathfrak{F}$. The distribution of (S, T) at the ith level of $\mathfrak{F}$ is expressed by

$$p(S = j, T = k; \mathfrak{F} = i) = p_{ijk}$$

$$(i = 1,2, \ldots, r; j = 1,2, \ldots, s; k = 1,2, \ldots, t).$$

A sample of size n_{i00} is taken from the population corresponding to $\mathfrak{F} = i$, and n_{ijk} is the frequency of the event $(S = j, T = k)$. For each i, the variates $\{N_{ijk}\}$ have a multinomial distribution with index n_{i00} and probabilities $\{p_{ijk}\}$. A transformation to the parameters introduced in section 7.1 shows that the distribution of $\{N_{ijk}\}$ is

$$\Pi_{ijk} (n_{i00}!\, p_{ist}^{n_{i00}}/n_{ijk}!) \exp (\Sigma_b\, n_{0b0}\phi_b + \Sigma_c\, n_{00c}\psi_c$$
$$+ \Sigma_{ab}\, n_{ab0}\lambda_{ab} + \Sigma_{ac}\, n_{a0c}\mu_{ac} + \Sigma_{bc}\, n_{0bc}\nu_{bc} + \Sigma_{abc}\, n_{abc}\xi_{abc}).$$

The parameters in the exponential term can be interpreted as follows. Consider the analysis of variance of a one-way layout with expected values $\{\eta_i\}$. The main effect is usually defined by the set of differences $\{\eta_i - \eta_0/r\}$ for $i = 1, 2, \ldots, r$, but an equally valid definition is given by the set of differences $\{\eta_a - \eta_r\}$ for $a = 1, 2, \ldots, r - 1$. At the ith level of $\mathfrak{F}$, the interaction $S \times T$ is measured by the parameters

$$\log (p_{ibc}p_{ist}/p_{ibt}p_{isc}) \quad \text{for all } b \text{ and } c.$$

On replacing η_i by $\log (p_{ibc}p_{ist}/p_{ibt}p_{isc})$, we find that the parameters $\{\xi_{abc}\}$ measure the effect of $\mathfrak{F}$ on $S \times T$. If H_ξ holds, then $S \times T$ is the same at each level of $\mathfrak{F}$ and is measured by $\{\nu_{bc}\}$. When H_ν holds, we can proceed to consider the effects of $\mathfrak{F}$ on S and T individually, which are measured by $\{\lambda_{ab}\}$ and $\{\mu_{ac}\}$ respectively. If H_ξ, H_ν, H_μ and H_λ all hold, then S and T are independent and each has the same distribution for all levels of $\mathfrak{F}$.

Secondly, suppose that a response T is observed for each treatment combination of two factors $\mathfrak{F}$ and $\mathfrak{G}$. The distribution of T for the treatment combination $(\mathfrak{F} = i, \mathfrak{G} = j)$ is expressed by

$$p(T = k; \mathfrak{F} = i, \mathfrak{G} = j) = p_{ijk}$$

$$(i = 1,2, \ldots, r; j = 1,2, \ldots, s; k = 1,2, \ldots, t).$$

A sample of size n_{ij0} is taken from the population corresponding to $(\mathfrak{F} = i, \mathfrak{G} = j)$, and n_{ijk} is the frequency of the event $T = k$. For each (i, j), the variates $\{N_{ijk}\}$ have a multinomial distribution with index n_{ij0} and probabilities $\{p_{ijk}\}$. Hence the distribution is also

$$\Pi_{ijk} (n_{ij0}!\, p_{ijt}^{n_{ij0}}/n_{ijk}!) \exp (\Sigma_c\, n_{00c}\psi_c + \Sigma_{ac}\, n_{a0c}\mu_{ac}$$
$$+ \Sigma_{bc}\, n_{0bc}\nu_{bc} + \Sigma_{abc}\, n_{abc}\xi_{abc}).$$

The interpretation of the parameters in the exponential term is now as follows. Consider the analysis of variance of a two-way layout with expected values $\{\eta_{ij}\}$. Instead of the usual definitions of main effects and interaction, we define the main effect of $\mathfrak{F}$ by $\{\eta_{as} - \eta_{rs}\}$ for $a = 1, 2, \ldots, r-1$, the main effect of $\mathfrak{G}$ by $\{\eta_{rb} - \eta_{rs}\}$ for $b = 1, 2, \ldots, s-1$, and the interaction $\mathfrak{F} \times \mathfrak{G}$ by

$$\eta_{ab} - \eta_{as} - \eta_{rb} + \eta_{rs} \quad \text{for all } a \text{ and } b.$$

For the treatment combination ($\mathfrak{F} = i$, $\mathfrak{G} = j$), the main effect of T is measured by

$$\log(p_{ijc}/p_{ijt}) \quad \text{for all } c.$$

On replacing η_{ij} by $\log(p_{ijc}/p_{ijt})$, we find that the parameters $\{\xi_{abc}\}$ measure the effect of $\mathfrak{F} \times \mathfrak{G}$ on T. When H_ξ holds, the effect of $\mathfrak{G}$ on T is the same at each level of $\mathfrak{F}$ and is measured by $\{\nu_{bc}\}$. Similarly, the effect of $\mathfrak{F}$ on T is the same at each level of $\mathfrak{G}$, and is measured by $\{\mu_{ac}\}$. If H_ξ, H_ν and H_μ all hold, then the distribution of T is the same for each treatment combination.

The conditional distributions which have already been derived can therefore be used to make inferences when a bivariate response is observed at each level of a factor, or when a response is observed for each treatment combination of two factors. Further details of the second experimental situation when $s = t = 2$ are given by Zelen (1971).

All these techniques arise from the measurement of association by the logarithm of the cross-product ratio. However, the arguments in favour of this measure no longer apply when a response is observed for each treatment combination of two factors because the table is not then symmetrical in the classifications. Suppose that the response is binary, and write p for the probability of success. The relationship between response and treatment combination can be expressed by a linear model for some function $f(p)$. Functions in common use are p itself, and the probit, angular and logit transformations (Cox, 1970a, sect. 2.7; Finney, 1971, sect. 3.8). For example, consider an experiment where $\mathfrak{F}$ has 2 levels and $\mathfrak{G}$ has s levels. The difference between the levels of $\mathfrak{F}$ at the jth level of $\mathfrak{G}$ is then measured by

$$\delta_j = f(p_{1j1}) - f(p_{2j1}).$$

Maximum-likelihood procedures are applied to this situation by Yates (1955a). When $\delta_j = \delta$, a locally most powerful criterion for testing the hypothesis $\delta = 0$ against the alternative $\delta \neq 0$ is based on a weighted sum of the differences

$$N_{1j1}/n_{1j0} - N_{2j1}/n_{2j0},$$

where the weights depend on the transformation used (Radhakrishna, 1965).

7.5 Estimation and test procedures

Write $\boldsymbol{\xi}$ for the vector obtained when the parameters $\{\xi_{abc}\}$ form a column in systematic order. The conditional maximum-likelihood estimator $\tilde{\boldsymbol{\xi}}$ satisfies the system of equations

$$N_{abc} = E(N_{abc}; \tilde{\boldsymbol{\xi}}) \quad \text{for all } a, b, \text{ and } c.$$

A solution can be obtained, in principle, by further generalization of the method for calculating $\tilde{\boldsymbol{\lambda}}$ outlined in section 5.2. In practice, the computational difficulties are likely to be severe.

The unconditional maximum-likelihood estimator of ξ_{abc} is

$$\hat{\xi}_{abc} = \log(N_{abc}N_{ast}N_{rbt}N_{rsc}/N_{abt}N_{asc}N_{rbc}N_{rst}).$$

As in section 5.2, the estimator

$$\xi^+_{abc} = \log\left\{\frac{(N_{abc}+\frac{1}{2})(N_{ast}+\frac{1}{2})(N_{rbt}+\frac{1}{2})(N_{rsc}+\frac{1}{2})}{(N_{abt}+\frac{1}{2})(N_{asc}+\frac{1}{2})(N_{rbc}+\frac{1}{2})(N_{rst}+\frac{1}{2})}\right\}$$

has a finite expectation and its bias is small. Similarly, we find that

$$Z^2 = (\boldsymbol{\xi}^+ - \boldsymbol{\xi})^T[\Delta\{N_{ijk}+\tfrac{1}{2}\}]^{-1}(\boldsymbol{\xi}^+ - \boldsymbol{\xi})$$

converges in distribution to $\chi^2_{(r-1)(s-1)(t-1)}$, where the dispersion matrix $\Delta\{n_{ijk}+\frac{1}{2}\}$ is calculated by treating the variates $\{\log(N_{ijk}+\frac{1}{2})\}$ as if they were mutually independent and such that

$$\text{var}\{\log(N_{ijk}+\tfrac{1}{2})\} = 1/(n_{ijk}+\tfrac{1}{2}).$$

The statistic Z^2 leads at once to a confidence region for $\boldsymbol{\xi}$ and simultaneous intervals for $\{\xi_{abc}\}$.

We next consider tests of H_ξ. The basis of an exact test is the distribution of $\{N_{abc}\}$ conditional on the observed two-dimensional marginal totals. However, the number of possible sets of $\{N_{abc}\}$ escalates rapidly with the value of $(r-1)(s-1)(t-1)$, and consequently there are usually serious difficulties in assessing the significance of the table observed. In general, a test of H_ξ is bound to involve approximate methods which are valid in large samples. As before, the standard test criteria are

$$X^2 = \Sigma_{ijk}\{N_{ijk} - m_{ijk}(\mathbf{0})\}^2/m_{ijk}(\mathbf{0}),$$

and

$$Y^2 = 2\,\Sigma_{ijk} N_{ijk}\log\{N_{ijk}/m_{ijk}(\mathbf{0})\}.$$

When H_ξ is true, X^2 converges in distribution to $\chi^2_{(r-1)(s-1)(t-1)}$ under the conditions stated in section 7.3 for the asymptotic normality of $\{N_{abc}\}$. A generalization of arguments given in sections 4.5 and 5.2 shows that $m_{ijk}(\mathbf{0})$ is the maximum-likelihood estimate of np_{ijk} under H_ξ, and that the likelihood-ratio statistic Y^2 is asymptotically equivalent to X^2.

The expected frequencies $\{m_{ijk}(\mathbf{0})\}$ are calculated by the iterative scaling procedure. We begin with any table which satisfies H_ξ, e.g. a table in which every frequency is 1, and continue by scaling the totals over i, j, and k in turn until all the observed two-dimensional marginal totals $\{n_{0jk}\}$, $\{n_{i0k}\}$, and $\{n_{ij0}\}$ are attained to the accuracy specified. An algorithm for calculating the estimated values of expected frequencies is given by Haberman (1972).

When $r = 2$, H_ξ implies the linear logistic model

$$\log(p_{1jk}/p_{2jk}) = \mu + \alpha_j + \beta_k .$$

The techniques of minimum logit chi-square can then be used to test H_ξ (Berkson, 1968) and there is evidence that in small samples they perform more successfully than X^2 or Y^2 (Odoroff, 1970).

If H_ξ is rejected, the reason may be that the parameters $\{\xi_{abc}\}$ can be expressed in terms of a smaller number of parameters. In that case, we proceed by analogy with section 6.2. When there is no systematic pattern among the $\{\xi_{abc}\}$, we must subdivide the three-dimensional table into constituent sets of two-dimensional tables, and analyse these separately by the methods of the three previous chapters.

Suppose that H_ξ is accepted. Inferences about $\{\nu_{bc}\}$ are based on the distribution of $\{N_{0bc}\}$ conditional on the observed values of $\{N_{ij0}\}$ and $\{N_{i0k}\}$, which is a convolution of r extended hypergeometric distributions. Write $\boldsymbol{\nu}$ for the vector defined by a systematic ordering of the parameters $\{\nu_{bc}\}$. The conditional maximum-likelihood estimator $\tilde{\boldsymbol{\nu}}$ satisfies the system of equations

$$N_{0bc} = E(N_{0bc}; \tilde{\boldsymbol{\nu}}) \quad \text{for all } b \text{ and } c.$$

When $s = t = 2$ some progress has been made. Approximations to $\tilde{\nu}_{11}$ are available when ν_{11} is near to zero (Birch, 1964; Goodman, 1969), and an exact treatment is possible when the fixed marginal totals are very small (Gart, 1970).

A large-sample approach to inferences about ν when H_ξ holds is as follows. For specified values of $\{\nu_{bc}\}$ let $\{m_{ijk}(\mathbf{0}, \nu)\}$ be the unique quantities which satisfy

$$\Sigma_j\, m_{ijk}(\mathbf{0},\boldsymbol{\nu}) = n_{i0k}, \quad \Sigma_k\, m_{ijk}(\mathbf{0},\boldsymbol{\nu}) = n_{ij0},$$

and

$$\log\{m_{ibc}(\mathbf{0},\boldsymbol{\nu})m_{ist}(\mathbf{0},\boldsymbol{\nu})/m_{ibt}(\mathbf{0},\boldsymbol{\nu})m_{isc}(\mathbf{0},\boldsymbol{\nu})\} = \nu_{bc}.$$

Then the likelihood-ratio statistic Y^2, given by

$$Y^2 = 2\,\Sigma_{ijk}\, N_{ijk} \log\{m_{ijk}(\mathbf{0})/m_{ijk}(\mathbf{0},\boldsymbol{\nu})\},$$

converges in distribution to $\chi^2_{(s-1)(t-1)}$ when $n_{i0k} \to \infty$ and $n_{ij0} \to \infty$ such that n_{i0k}/n and n_{ij0}/n have positive limits for all i, j, and k. Consider the identity

$$\Sigma_{ijk}\{n_{ijk} - m_{ijk}(\boldsymbol{\xi})\}\log m_{ijk}(\boldsymbol{\xi}) - \Sigma_{abc}\{n_{abc} - m_{abc}(\boldsymbol{\xi})\}\xi_{abc} \equiv 0.$$

By taking $\boldsymbol{\xi} = \mathbf{0}$, we obtain

$$\Sigma_{ijk}\{N_{ijk} - m_{ijk}(\mathbf{0})\}\log m_{ijk}(\mathbf{0}) = 0.$$

The replacement of $\log m_{ijk}(\boldsymbol{\xi})$ by $\log m_{ijk}(\mathbf{0},\boldsymbol{\nu})$ further shows that

$$\Sigma_{ijk}\{N_{ijk} - m_{ijk}(\mathbf{0})\}\log m_{ijk}(\mathbf{0},\boldsymbol{\nu}) = 0.$$

We can therefore write Y^2 in the form

$$Y^2 = 2\,\Sigma_{ijk}\, m_{ijk}(\mathbf{0})\log\{m_{ijk}(\mathbf{0})/m_{ijk}(\mathbf{0},\boldsymbol{\nu})\}.$$

An asymptotically equivalent statistic is

$$X^2 = \Sigma_{ijk}\,\{(m_{ijk}(\mathbf{0}) - m_{ijk}(\mathbf{0},\boldsymbol{\nu})\}^2/m_{ijk}(\mathbf{0},\boldsymbol{\nu}).$$

When the conditions for convergence in distribution of X^2 and Y^2 are suitably approximated, a test of H_ν assuming that H_ξ holds is made using either statistic with the expected frequencies

$$m_{ijk}(\mathbf{0},\mathbf{0}) = n_{ijo}\,n_{iok}/n_{ioo}.$$

Another test procedure is the following (Birch, 1965). We require to test the simple hypothesis that $\nu_{bc} = 0$ for all b and c against the composite alternative that $\nu_{bc} \neq 0$ for at least one pair (b, c). An appropriate statistic is S_3 of section 2.3, denoted here by K^2. Under H_ξ, the derivatives of the conditional log likelihood function with respect to $\{\nu_{bc}\}$ at $\{\nu_{bc} = 0\}$ form a vector $\mathbf{h}$ with elements typified by

$$h_{bc} = \Sigma_i\,(N_{ibc} - n_{ibo}\,n_{ioc}/n_{ioo}).$$

When H_ξ and H_ν both hold, the exact means, variances, and covariances of $\{N_{ijk}\}$ conditional on the observed values of $\{N_{ijo}\}$ and $\{N_{iok}\}$ are given by

$$E\,(N_{ijk}) = n_{ijo}\,n_{iok}/n_{ioo},$$

and

$$\operatorname{cov}(N_{ijk}, N_{ij'k'}) = n_{ijo}(\delta_{jj'}n_{ioo} - n_{ij'o})n_{iok}(\delta_{kk'}n_{ioo} - n_{iok'})/n_{ioo}^2(n_{ioo} - 1).$$

Let $\mathbf{D}$ be the dispersion matrix of $\mathbf{h}$, with elements typified by $\Sigma_i \operatorname{cov}(N_{ibc}, N_{ib'c'})$. Then

$$K^2 = \mathbf{h}^T\mathbf{D}^{-1}\mathbf{h}.$$

The asymptotic distribution of K^2 is $\chi^2_{(s-1)(t-1)}$, not only under the conditions which apply to X^2 and Y^2, but also when $r \to \infty$, in which case the marginal totals $\{n_{ijo}\}$ and $\{n_{iok}\}$ may be small. On the other hand, K^2 is generally more difficult to calculate than X^2 or Y^2.

The standard method of deriving confidence regions from test statistics, described in section 5.2, can be applied to either X^2 or Y^2 in respect of $\boldsymbol{\nu}$, but the problems of enumeration are considerable when $(s-1)(t-1)>2$. When $s=t=2$, the method presented by Cornfield (1956) and described in section 4.4 extends directly to give confidence intervals for ν_{11} (Gart, 1970).

Suppose that H_ν is accepted in addition to H_ξ. Estimation and test procedures for $\{\mu_{ac}\}$ and $\{\lambda_{ab}\}$ are based on a single extended hypergeometric distribution for each set of parameters. The relevant methods are discussed in previous chapters. If H_μ and H_λ are also accepted, then R, S and T are mutually independent.

Consider the identity

$$p(N_{ijk}|N_{i00}, N_{0j0}, N_{00k})$$
$$\equiv p(N_{ijk}|N_{ij0}, N_{i0k}, N_{0jk})p(N_{0jk}|N_{ij0}, N_{i0k})p(N_{i0k}|N_{ij0}, N_{00k}) \times p(N_{ij0}|N_{i00}, N_{0j0}, N_{00k}).$$

Successive terms on the right give the distributions for inferences about (i) $\{\xi_{abc}\}$, (ii) $\{\nu_{bc}\}$ when $\{\xi_{abc}=0\}$, (iii) $\{\mu_{ac}\}$ when $\{\xi_{abc}=0\}$ and $\{\nu_{bc}=0\}$, and (iv) $\{\lambda_{ab}\}$ when $\{\xi_{abc}=0\}$, $\{\nu_{bc}=0\}$ and $\{\mu_{ac}=0\}$. A similar result holds when each conditional probability $p(.)$ is replaced by a likelihood ratio $l(.)$. The corresponding identity in terms of $-2\log l(.)$ shows that the likelihood-ratio statistic for testing that R, S, and T are mutually independent is the sum of four such quantities which have asymptotic χ^2 distributions and are mutually independent when all the relevant null hypotheses are true (Goodman, 1969, 1970).

Each of the likelihood-ratio statistics in the foregoing analysis can be further partitioned. For example, suppose that H_ξ holds, and we require to partition the statistic for testing H_ν. Introduce the hypothesis

$$J_b : \nu_{bc} = 0 \quad \text{for all } c.$$

Then a possible partition consists of the likelihood-ratio statistics for testing the sequence of hypotheses formed by J_1, J_2 when J_1 holds, J_3 when J_1 and J_2 hold, . . . , and J_{s-1} when $J_1, J_2, \ldots, J_{s-2}$ hold. The expected frequencies corresponding to J_1 are obtained from any table which satisfies H_ξ by scaling until the marginal totals $\{n_{ij0}\}$, $\{n_{i0k}\}$ and $\{n_{0bc}\}$ for $b \neq 1$ agree satisfactorily with the values observed. Similarly, the expected frequencies corresponding to J_2 when J_1 holds are derived by fixing the marginal totals $\{n_{ij0}\}$, $\{n_{i0k}\}$, and $\{n_{0bc}\}$ for $b \neq 1, 2$. The procedure terminates with J_{s-1} when the restrictions on $\{n_{0bc}\}$ disappear and the expected frequencies are

$$m_{ijk}(\mathbf{0}, \mathbf{0}) = n_{ij0}\, n_{i0k}/n_{i00}.$$

By this method, the $(s-1)(t-1)$ degrees of freedom for H_ν are divided into $(s-1)$ sets of $(t-1)$. They could equally be divided into $(t-1)$ sets of $(s-1)$. Any other method of grouping the elementary hypotheses typified by $\nu_{bc} = 0$ is unlikely to have a practical interpretation, and the calculation of the corresponding expected frequencies will present greater difficulties.

7.6 Numerical examples

We proceed to analyse sets of data which are described in section 7.1.

Example 7.5 The data are given in Table 7.1. We use the following notation.

Length of cutting ($\mathfrak{F}$): long/short ($i = 1, 2$).

Time of planting ($\mathfrak{G}$): at once/in spring ($j = 1, 2$).

Response (T): alive/dead ($k = 1, 2$).

Consider first the hypothesis H_ξ, according to which the interaction between length of cutting and time of planting has no effect on response. The expected frequencies $\{m_{ijk}(0)\}$ are calculated by the iterative scaling procedure, beginning with a table in which all frequencies are 1, and scaling to achieve the two-dimensional marginal totals given below.

Cell	$\mathfrak{F}\mathfrak{G}$	$\mathfrak{F}T$	$\mathfrak{G}T$
(1, 1)	240	240	263
(1, 2)	240	240	217
(2, 1)	240	138	115
(2, 2)	240	342	365

Denote n_{111} by a. The following table expresses the other frequencies in terms of a under the conditions on the marginal totals, and gives n_{ijk} and $m_{ijk}(0)$.

$\mathfrak{F}\mathfrak{G}T$	Frequency	n_{ijk}	$m_{ijk}(0)$	$\mathfrak{F}\mathfrak{G}T$	Frequency	n_{ijk}	$m_{ijk}(0)$
(111)	a	156	161·096	(211)	$263-a$	107	101·904
(112)	$240-a$	84	78·904	(212)	$a-23$	133	138·096
(121)	$240-a$	84	78·904	(221)	$a-125$	31	36·096
(122)	a	156	161·096	(222)	$365-a$	209	203·904

Under H_ξ, the frequency A is approximately normal with mean 161·096 and standard deviation

$$(161{\cdot}096^{-1} + 78{\cdot}904^{-1} + \ldots + 203{\cdot}904^{-1})^{-\frac{1}{2}} = 3{\cdot}382.$$

We require the probability of a deviation from the mean in either direction as large or larger than that observed, namely

$$p(A \leqslant 156) + p(A \geqslant 167).$$

The normal approximation with a continuity correction gives

$$0{\cdot}087 + 0{\cdot}055 = 0{\cdot}142.$$

We conclude that $\mathfrak{F} \times \mathfrak{G}$ has no effect on T.

An exact test of H_ξ is made using the conditional distribution of A, which is given by

$$p(A = a) = c/[(a!)^2\{(240-a)!\}^2(263-a)!\,(a-23)!\,(a-125)!\,(365-a)!]$$

Here c is chosen so that the total probability is 1. The mean and standard deviation of A imply that all frequencies outside the range 161 ± 12 have negligible probabilities. Thus c can be evaluated. The probability of observing $A = 156$ or a less probable frequency is again 0·142.

Consider next the effect of $\mathfrak{G}$, say ν. The 2×2 tables for $\mathfrak{F} = i$ have the same cross-product ratio e^ν. For each i and values of ν we calculate the expected frequency $m_{i11}(0, \nu)$ and variance

$$\{m_{i11}(0,\nu)\}^{-1} + \{m_{i12}(0,\nu)\}^{-1} + \{m_{i21}(0,\nu)\}^{-1} + \{m_{i22}(0,\nu)\}^{-1}.$$

The conditional mean and variance of N_{011} are obtained by summation over i. This leads to two values of ν, such that 263 appears at the upper and lower $2\frac{1}{2}$ percent points of the distribution of N_{011}, assumed normal. In this way we find that the effect of planting at once rather than in spring lies in the interval (1·130, 1·725), with a confidence coefficient approximately 0·95. Similarly, the effect of $\mathfrak{F}$ lies in (0·723, 1·313), with the same confidence coefficient.

Example 7.6 The data are given in Table 7.2. We use the following notation:

Birth-order ($\mathfrak{F}$): 2/3–4/5+ $(i = 1, 2, 3)$.

Group ($\mathfrak{G}$): problems/controls $(j = 1, 2)$.

Response (T): losses/none $(k = 1, 2)$.

Consider first the hypothesis H_ξ. The expected frequencies $\{m_{ijk}(\mathbf{0})\}$ are given in Table 7.5. We find that $Y^2 = 0{\cdot}853$ (2 d.f.), and conclude that $\mathfrak{F} \times \mathfrak{G}$ has no effect on T.

Consider next the hypothesis H_ν. The expected frequencies $m_{ijk}(\mathbf{0}, \mathbf{0})$ are given in Table 7.5. Here $Y^2 = 2{\cdot}301$ (1 d.f.), and we conclude that $\mathfrak{G}$ has no effect on T. An alternative test is based on the conditional distribution of N_{011}. Under H_ν, the means and variances of $\{N_{i11}\}$ are as follows:

	N_{111}	N_{211}	N_{311}
Mean	18·434	24·903	23·360
Variance	5·858	6·426	5·321

TABLE 7.5 *Losses of mothers in relation to behavioural problems and birth-order of children: expected frequencies $m_{ijk}(0)$ and $m_{ijk}(0, 0)$, respectively upper and lower*

Number of mothers with	Birth-order 2		3–4		5+	
	Problems	Controls	Problems	Controls	Problems	Controls
Losses	20·503 18·434	9·497 11·566	27·213 24·903	14·787 17·097	25·284 23·360	15·716 17·640
None	81·497 83·566	54·503 52·434	39·787 42·097	31·213 28·903	23·716 25·640	21·284 19·360

Hence N_{011} has mean 66·70, variance 17·60, and standard deviation 4·20. The probability of a deviation as large or larger in the same direction is 0·084.

Finally, we evaluate the effect of $\mathfrak{F}$. Let μ_1 define the change in response from a birth-order of 2 to one of 3–4, and let μ_2 define the change from 2 to 5+. The 3 X 2 tables for $\mathfrak{G} = j$ have the same pair of cross-product ratios (e^{μ_1}, e^{μ_2}). For each pair (μ_1, μ_2), we calculate the expected frequencies $\{m_{ijk}(\mathbf{0}, \boldsymbol{\mu})\}$, and then

$$Y^2 = 2 \Sigma_{ijk} N_{ijk} \log \{m_{ijk}(\mathbf{0})/m_{ijk}(\mathbf{0}, \boldsymbol{\mu})\}.$$

By scanning the plane of (μ_1, μ_2), we obtain all pairs such that Y^2 lies at specified percentage levels of χ^2_2. The resulting confidence distribution is shown in Fig. 7.1.

Example 7.7 The data are given in Table 7.4. We use the following notation:

Breathlessness (R): Yes/No ($i = 1, 2$).

Wheeze (S): Yes/No ($j = 1, 2$).

Age (T): 20–24/. . ./60–64 ($k = 1, 2, \ldots, 9$).

A test of H_ξ gives $Y^2 = 26{\cdot}686$ on 8 d.f. ($P < 0{\cdot}001$). The log cross-product ratios in order of increasing age are as follows:

3·21 3·69 3·40 3·14 3·01 2·78 2·92 2·44 2·64

They conform roughly to a linear trend with a slope of $-0{\cdot}15$. We therefore consider the hypothesis that

$$\xi_{11c} = (c-1)\beta \quad \text{for some } \beta.$$

For values of β in the neighbourhood of $-0{\cdot}15$, a table of expected frequencies $\{m_{ijk}(\beta)\}$ is calculated, beginning with the table below and using

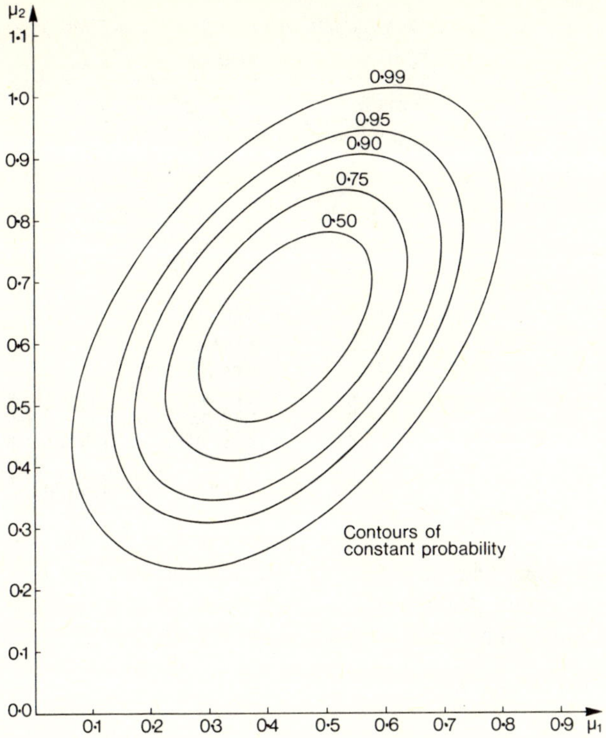

Fig. 7.1 *Confidence distribution for effect of factor $\mathcal{F}$ in 3 × 2 × 2 contingency table*

the iterative scaling procedure to achieve the observed two-dimensional marginal totals.

			Age-groups			
No breathlessness	No wheeze	1	e^{β}	$e^{2\beta}$	...	$e^{8\beta}$
	Wheeze	1	1	1	...	1
Breathlessness	No wheeze	1	1	1	...	1
	Wheeze	1	1	1	...	1

The minimum value of Y^2 is 6·805 (7 d.f.), which indicates a satisfactory fit. This value is attained when $\beta = -0{\cdot}1306$, and the expected frequencies are then as shown in Table 7.6. A confidence interval for β with a confidence coefficient of 0·95 consists of all the values for which

$$Y^2 \leqslant 6{\cdot}805 + 3{\cdot}841 = 10{\cdot}646,$$

where 3·841 is the upper 5 percent point of χ_1^2. The resulting interval extends from −0·1887 to −0·0731.

TABLE 7.6 *Expected frequencies of coalminers for linear regression of log cross-product ratio on age*

Age-group in years	Breathlessness		No breathlessness		Total
	Wheeze	No wheeze	Wheeze	No wheeze	
20–24	10·21	5·79	93·79	1842·21	1952·00
25–29	21·21	10·79	106·79	1652·21	1791·00
30–34	52·49	20·51	178·51	1861·49	2113·00
35–39	121·39	47·61	256·61	2357·39	2783·00
40–44	169·29	53·71	272·71	1778·29	2274·00
45–49	274·77	82·23	318·23	1717·77	2393·00
50–54	393·34	127·66	255·66	1313·34	2090·00
55–59	418·79	139·21	212·21	979·79	1750·00
60–64	365·51	112·49	138·49	519·51	1136·00
Total	1827·00	600·00	1833·00	14 022·00	18 282·00

This analysis deals only with the question of how the association between breathlessness and wheeze varies with age, and discusses that question from only one viewpoint. Many other matters remain for consideration. They are examined in detail by Ashford and Sowden (1970), and Mantel and Brown (1973).

8 MATCHING

8.1 Designs and hypotheses

Suppose that s treatments are compared by observing a response obtained from experimental subjects. Matching is used in an attempt to improve the precision of comparisons between treatments from the level attained by random samples. The procedure is to classify rs subjects into r groups of size s by means of a set of matching variates which are correlated with each other and with the degree of response. Subjects in the same group are regarded as equivalent with respect to the matching variates, and a different treatment is applied to each subject. The experimental design is known as *matched pairs* when $s = 2$, and as *matched sets* in general.

When the response has t categories, the experimental data are the frequencies in a contingency table of order $r \times s \times t$, where the classifications refer to groups, treatments, and response, respectively. The sample size is 1 for each combination of group and treatment. Methods of presenting the data are illustrated in the examples which follow. The general principles of matching are discussed from a practical viewpoint by Billewicz (1965).

Example 8.1 The data in Table 8.1 are quoted from Maxwell (1961). They were obtained in an experiment concerned with the effect of the symptom

TABLE 8.1 *Recovery of 23 pairs of depressed patients*

Not depersonalized	Depersonalized: Recovered	Depersonalized: Not recovered	Total
Recovered	14	5	19
Not recovered	2	2	4
Total	16	7	23

"depersonalization" on the prognosis of depressed patients. A set of 23 "depersonalized" patients was matched one-to-one for age, sex, duration of illness, and on certain personality variables, with 23 patients diagnosed as "not depersonalized". After a course of E.C.T. the members of each pair were diagnosed as "recovered" or "not recovered" and the corresponding frequencies are given in the table. We require to test the hypothesis that "depersonalization" is not associated with prognosis against the alternative that the categories "not depersonalized" and "recovered" are positively associated. Here $r = 23$, $s = 2$, and $t = 2$. Although Table 8.1 is formally equivalent to a 2×2 contingency table, the units are pairs of subjects, and the usual methods do not apply.

Example 8.2 The data in Table 8.2 are quoted from Cochran (1950). Specimens of diphtheria bacilli were taken from the throats of 69 suspected cases.

TABLE 8.2 *Growth of 69 specimens of diphtheria bacilli*

Medium				Frequency
A	*B*	*C*	*D*	
1	1	1	1	4
1	1	0	1	2
0	1	1	1	3
0	1	0	1	1
0	0	0	0	59

Each specimen was grown on each of four media A, B, C, D. The result was either success, counted as 1, or failure, counted as 0. We require to test that the different media are equally effective, taking account of the fact that the probability of growth will depend on the number of diphtheria bacilli present. Here $r = 69$, $s = 4$, and $t = 2$.

Most applications of matching to qualitative variates are concerned with a binary response, and we therefore make the assumption that $t = 2$ throughout what follows. The methods of analysis are largely obtained by specializing previous results and can be extended to other values of t without difficulty. An appropriate model for matched sets is described by Miettinen (1968). Matching is based on a variate M, which is usually multidimensional. The value of M for the ith group is m_i, where $m_1, m_2, \ldots, m_r$ form a random sample of size r. For each value of M, observations are made on the binary variates $T_1, T_2, \ldots, T_s$, where T_j refers to the jth treatment. Denote the variate $p(T_j = 1|M)$ by P_j, and its expectation by p_j. Thus

$$p(T_j = 1) = p_j.$$

We assume that $T_1, T_2, \ldots, T_s$ are mutually independent conditional on $M = m$, in the following sense. Let $T_j(m)$ be the variate corresponding to the distribution of T_j conditional on $M = m$. Then $T_1(m), T_2(m), \ldots, T_s(m)$ are independent for all m. This assumption implies that

$$p(T_1 = u, T_2 = v, \ldots, T_s = w) = E\{p(T_1 = u|M)p(T_2 = v|M) \ldots p(T_s = w|M)\}$$

for all $u, v, \ldots, w$.

There are several ways in which the condition that the treatments are identical can be expressed (Madansky, 1963). A strong formulation is given by the hypothesis H_0, according to which the conditional probability of success is the same for all possible values of m. Thus

$$p(T_1 = 1|m) = p(T_2 = 1|m) = \ldots = p(T_s = 1|m) \quad \text{for all } m.$$

The variates $P_1, P_2, \ldots, P_s$ are then identical. Write

$$\theta_{uv\ldots w} = p(T_1 = u, T_2 = v, \ldots, T_s = w).$$

The probabilities $\{\theta_{uv\ldots w}\}$ can be located in the cells of a contingency table of order 2^s, where the jth classification refers to the response for the jth treatment. A consequence of H_0 is the hypothesis H_I, according to which $\theta_{uv\ldots w}$ is invariant under permutations of the suffixes. This invariance can be expressed by the condition that the d-dimensional margins of the contingency table are homogeneous for $d = 1, 2, \ldots, s-1$. Consider the hypothesis

$$H : p_1 = p_2 = \ldots = p_s.$$

This is the condition that the one-dimensional margins are homogeneous. Thus H_I implies H, but H does not imply H_I except when $s = 2$. For example, take $s = 3$. Then H_I is given by

$$\theta_{100} = \theta_{010} = \theta_{001} \quad \text{and} \quad \theta_{011} = \theta_{101} = \theta_{110}.$$

On the other hand, H is given by

$$\theta_{100} - \theta_{011} = \theta_{010} - \theta_{101} = \theta_{001} - \theta_{110}.$$

We say that the grouping of subjects is *ineffective* if $T_1, T_2, \ldots, T_s$, and M are mutually independent, in which case

$$p(T_j = 1|m) = p_j \quad \text{for all } j \text{ and } m.$$

All these expressions have counterparts which are conditional on the values of M observed. Denote by T_{ij} the variate defined by the distribution of T_j conditional on $M = m_i$. Write

$$p_{ij} = p(T_{ij} = 1) \quad \text{and} \quad q_{ij} = p(T_{ij} = 0).$$

The natural analogue of H_0, also denoted by H_0, is expressed by

$$p_{i1} = p_{i2} = \ldots = p_{is} \quad \text{for all } i.$$

However, the analogue of H asserts that the average probability of success for the subjects in the sample is the same for all treatments, and this statement has limited value. The grouping is ineffective when

$$p_{ij} = p_j \quad \text{for all } i \text{ and } j.$$

This condition obtains when s independent binomial samples are matched at random.

Matching is a particular form of stratification, where the procedure is as follows. We classify the subjects into homogeneous samples, each of size larger than s, and then apply the different treatments to sub-samples which are generally unequal in size. In other words, the sample size for each combination of stratum and treatment, previously 1, is now arbitrary.

8.2 Matched sets

We begin with a matched pairs design. The basic assumption, which cannot be tested, is that the interaction of pairs and treatments has no effect on response. Thus H_ξ holds, and

$$\log (p_{i1}\, q_{i2}/p_{i2}q_{i1}) = \nu \quad \text{for all } i.$$

This shows that the probabilities of success for the two treatments, when measured on a logit scale, differ by the same amount for all pairs. According to the general theory of $r \times s \times t$ contingency tables, inferences about ν are based on the distribution of N_{011} conditional on the observed values of $\{N_{i0k}\}$. In the ith pair, the vector (n_{i01}, n_{i02}) has three possible values, namely (2, 0), (0, 2), and (1, 1). When the first value obtains, $N_{i11} = 1$; and when the second value obtains, $N_{i11} = 0$. Thus no variation is possible in either case, and no information about ν is available. Consider now the third possibility:

$$\begin{aligned} p(N_{i11} &= 1|N_{i01} = 1) \\ &= p_{i1}q_{i2}/(p_{i1}q_{i2} + p_{i2}q_{i1}) = (1 + e^{-\nu})^{-1} . \end{aligned}$$

The conditional probability here is the same for all pairs. Since $N_{i11} = 1$ implies $N_{i21} = 0$ when $N_{i01} = 1$, all the information about ν is derived from the pairs which give a different result for the two treatments. Such pairs are described as *untied*.

We are thus led to the following conclusions. Let T_{ij} denote the binary variate corresponding to the jth treatment in the ith pair, and let F_{uv} be the frequency of pairs for which $T_{i1} = u$ and $T_{i2} = v$. The frequency of untied pairs is $F_{10} + F_{01}$, say G, and its observed value is denoted by g. Inferences about ν are based on the distribution of F_{10} conditional on $F_{10} + F_{01} = g$, which is binomial with index g and probability $(1 + e^{-\nu})^{-1}$. When the treatments are identical, $\nu = 0$ and the binomial probability is $\frac{1}{2}$ .The procedure of referring F_{10} to a binomial distribution with index g and probability $\frac{1}{2}$ was pointed out by Mosteller (1952), and is formally equivalent to the sign test for a continuous variate with a symmetrical distribution and hypothetical mean zero. An approximate distribution function for F_{10} is obtained from $\mathfrak{N}(g/2, g/4)$, using a continuity correction. The corresponding test without a continuity correction is due to McNemar (1947), and consists in referring the statistic

$$X_I^2 = (F_{10} - F_{01})^2/(F_{10} + F_{01})$$

to a table of χ_1^2. In what follows, we refer to both the exact and asymptotic procedures as the sign test.

A notable feature of the sign test is that no use is made of the pairs which give the same result for each treatment. Suppose that the values of F_{10} and F_{01} are given. We might expect, as Cochran (1950, p. 258) has pointed out, that significance ought to be more definitely established if F_{00} and F_{11} are zero than if F_{00} and F_{11} are large. In fact, this property can follow from a Bayesian analysis where prior distributions are assigned to the treatment and pair effects. The details are given by Altham (1971a), where several Bayesian models appropriate for matched pairs are discussed.

Example 8.3 The data are given in Table 8.1. A one-sided test of the hypothesis that "depersonalization" is not associated with prognosis is made by calculating the probability of 2 or fewer successes in a sequence of 7 trials with probability $\frac{1}{2}$ for success. The probability is 29/128 and we conclude that "depersonalization" and prognosis are not associated.

When a matched pair is formed by applying treatments $\mathfrak{F}$ and $\mathfrak{G}$ to the same subject, we require to take account of a possible order effect. Suppose that one group of subjects is given the treatments in the order $(\mathfrak{F}, \mathfrak{G})$, and another group in the order $(\mathfrak{G}, \mathfrak{F})$. An appropriate analysis of the two values of ν was derived by Gart (1969), who showed that two applications of the Fisher–Irwin procedure provide the tests required. A numerical study of the relationship between the sign test and Gart's test was made by Nam (1971). He showed that the sign test is biased when an order effect exists, while Gart's test is less powerful than the sign test when the order effect is zero. The conclusion is that Gart's test should be used in place of the sign test when an order effect is suspected.

We now consider the analysis of a matched sets design when there are s treatments. The assumption is made again that H_ξ holds, so that

$$\log(p_{ib}q_{ib'}/p_{ib'}q_{ib}) = \nu_{b1} - \nu_{b'1} \quad \text{for all } i.$$

Thus the probabilities of success for any pair of treatments, when measured on a logit scale, differ by a constant amount for all sets. The hypothesis H_ν, according to which $\nu_{b1} = 0$ for all b, implies H_0, H_I and H. Inferences about $\{\nu_{b1}\}$ are based on the distribution of $\{N_{0b1}\}$ conditional on the observed values of $\{N_{i0k}\}$, first derived by Rasch (1960, ch. X). Consider the test of H_ν which uses the statistic K^2 of section 7.5. The frequency of success for the jth treatment is T_{0j}, and the frequency of success for the ith set is T_{i0}. Write

$$A = \Sigma T_{0j}/s,$$

so that A is the average number of successes per treatment. Then Birch (1965) showed that

$$K^2 = s(s-1)\,\Sigma\,(T_{0j}-A)^2/(s\,\Sigma\,t_{i0} - \Sigma\,t_{i0}^2),$$

where t_{i0} is the observed value of T_{i0}. This statistic was developed by Cochran (1950) as a standardized version of $\Sigma\,(T_{0j}-A)^2$. He denoted K^2 by Q and showed that Q converges in distribution to χ^2_{s-1} as $r \to \infty$ when H_I is true. An extension of Cochran's test statistic to the situation where results are missing for some combinations of group and treatment was made by Carothers (1971).

Example 8.4 The data are given in Table 8.2. They lead to the following frequency distributions:

t_{i0}	0	2	3	4
Frequency	59	1	5	4

t_{0j}	6	7	10
Frequency	1	1	2

Hence $Q = 8{\cdot}05$, near the upper 5 percent point of χ^2_3.

The exact distribution of Q can be derived, in principle at least, from the following considerations. Let $F_{uv\ldots w}$ be the number of sets for which

$$T_{i1} = u, \quad T_{i2} = v, \ldots, \quad T_{is} = w.$$

Write G_x for the number of sets such that $T_{i0} = x$. Thus G_x is the sum of $F_{uv\ldots w}$ over all permutations of the suffixes such that $t_{i0} = x$. Only one permutation is possible if $x = 0$ or s. When H_I is true,

$$p(T_{i1} = u, T_{i2} = v, \ldots, T_{is} = w | t_{i0} = x) = \binom{s}{x}^{-1}.$$

Thus the joint distribution of the frequencies $\{F_{uv\ldots w}\}$, conditional on $T_{i0} = x$ and $G_x = g_x$, is the symmetric multinomial

$$g_x! \Big/ \binom{s}{x}^{g_x} \Pi\, f_{uv\ldots w}!$$

The joint distribution of all frequencies $\{F_{uv\ldots w}\}$, conditional on $G_x = g_x$, is a product of such probabilities.

Bennett (1967, 1968) estimates the probabilities of the frequency distribution of $\{T_{i0}\}$ by various methods, subject to the constraints imposed by H, and his criterion is a goodness-of-fit statistic. The relative merits of the various test statistics are unknown.

8.3 Efficiency of matched pairs for binary data

The basic assumption for the matched pairs design is that

$$\log(p_{i1}q_{i2}/p_{i2}q_{i1}) = \nu \quad \text{for all } i.$$

Inferences about ν are made from the distribution of F_{10} conditional on $F_{10} + F_{01} = g$, which is binomial with index g and probability $(1 + e^{-\nu})^{-1}$. This method is valid whether or not the matching is effective. However, when the matching is ineffective, so that

$$p_{i1} = p_1 \quad \text{and} \quad p_{i2} = p_2 \quad \text{for all } i,$$

another method of making inferences about ν becomes available. Consider the following 2 X 2 table of treatment and response:

Treatment	Success	Failure
1	T_{01}	$r - T_{01}$
2	T_{02}	$r - T_{02}$

Here $T_{01} = F_{11} + F_{10}$ and $T_{02} = F_{11} + F_{01}$. Denote by t_0 the observed value of $T_0 = T_{01} + T_{02}$. Inferences about ν are now made from the distribution of T_{01}, conditional on $T_0 = t_0$, which has the extended hypergeometric form. We proceed to compare the two methods when matching is ineffective.

The hypothesis H, according to which $p_1 = p_2$, can be tested in large samples using either

$$X_I^2 = (F_{10} - F_{01})^2/g$$

or

$$X^2 = 2r(T_{01} - T_{02})^2/t_0(2r - t_0).$$

Each statistic converges in distribution to χ_1^2 when H is true. Wald (1947, p.108) remarked that the values of F_{10} and F_{01} depend on random pairing, which suggests that X_I^2 cannot be as efficient as X^2. However, he proved earlier (1943, App. A4) that the powers of X_I^2 and X^2 are asymptotically equal for a sequence of alternatives $\{(p_{1r}, p_{2r})\}$ such that the sequence $\{|p_{1r} - p_{2r}| \, r^{-\frac{1}{2}}\}$ is bounded. The power functions of the sign test and the Fisher–Irwin test have been calculated by Youkeles (1963) at illustrative values of (p_1, p_2) when r = 10, 15, 20 and 30. She found that the power of the sign test can be relatively low in small samples, but that the differences have virtually disappeared when r = 30. Chase (1968) considered alternatives of the type $p_2 = \xi p_1$, where $0 \leqslant \xi \leqslant 1$. He measured efficiency by the ratio of the sample size for X^2 to the expected sample size for X_I^2 when the number of untied pairs is fixed. The two statistics were compared under the same conditions of significance level, power, and alternative hypothesis. Three methods of comparison were considered, as follows.

(a) Fix the significance level α and power $1-\beta$. Consider a sequence of alternatives $p_2 = \xi_r p_1$ where $\xi_r \to 1$.

(b) Fix ξ and the power $1-\beta$. Consider a sequence of significance levels $\alpha_r \to 0$.

(c) Fix the significance level α. Consider related sequences $\{\xi_r\}$ and $\{\beta_r\}$ such that $\xi_r \to 1$ and $\beta_r \to 0$.

The conclusion is that the asymptotic efficiency of X_I^2 with respect to X^2 is 1 for (a) and (c), but less than 1 for (b). However, in most situations of practical interest, the asymptotic relative efficiency under (b) is quite high.

Another approach to the study of efficiency is to consider the problem as one of estimation. The maximum-likelihood estimator of ν from the conditional distribution of F_{10} is

$$\tilde{\nu}_I = \log F_{10} - \log F_{01}.$$

This estimator has infinite variance, but as $g \to \infty$ the distribution of

$$(\tilde{\nu}_I - \nu)\,\{\text{var}\,(\tilde{\nu}_I)\}^{-\frac{1}{2}}$$

converges to normality, where

$$\text{var}\,(\tilde{\nu}_I) = (1+e^{\nu})^2/ge^{\nu}.$$

Substituting for e^{ν}, and replacing g by $E(G)$, we obtain

$$\text{var}\,(\tilde{\nu}_I) = (p_1q_2 + p_2q_1)/rp_1p_2q_1q_2.$$

Denote by $\tilde{\nu}$ the maximum-likelihood estimator of ν from the conditional distribution of T_{01}. As $r \to \infty$,

$$\tilde{\nu} \sim \log T_{01} - \log\,(r - T_{01}) - \log T_{02} + \log\,(r - T_{02}).$$

With the same interpretation of var (.) as before,

$$\text{var}(\tilde{\nu}) = (1/p_1q_1 + 1/p_2q_2)/r.$$

A measure of asymptotic efficiency in the limit, as $r \to \infty$, is given by

$$\text{var}\,(\tilde{\nu})/\text{var}\,(\tilde{\nu}_I) = (p_1q_1 + p_2q_2)/(p_1q_2 + p_2q_1).$$

Now $p_1 - p_2$ and $q_1 - q_2$ are either both zero or different in sign, so that

$$(p_1 - p_2)(q_1 - q_2) \leqslant 0,$$

whence $$p_1q_1 + p_2q_2 \leqslant p_1q_2 + p_2q_1.$$

We conclude that

$$\text{var}\,(\tilde{\nu}) \leqslant \text{var}\,(\tilde{\nu}_I),$$

with equality if and only if $p_1 = p_2$, in which case $\nu = 0$. For a fixed value of p_2, say p, the relative efficiency tends to p as $\nu \to \infty$, and to $q = 1 - p$ as $\nu \to -\infty$.

When the matching is effective, $\tilde{\nu}_I$ remains a consistent estimator of ν, but $\tilde{\nu}$ is a consistent estimator of

$$\log(\Sigma\, p_{i1} \Sigma\, q_{i2} / \Sigma\, p_{i2}\, \Sigma\, q_{i1}).$$

Suppose that the values $\{p_{i1}\}$ form a random sample from the distribution of a variate P_1. We consider the special case where P_1 is uniform on (0, 1). Denote e^ν by ψ and define P_2 by

$$P_1(1-P_2)/P_2(1-P_1) = \psi.$$

Then $E(P_1) = \frac{1}{2}$ and

$$E(P_2) = \{(1-\psi) + \psi \log \psi\}/(1-\psi)^2.$$

In the limit, as $r \to \infty$, the bias of $\tilde{\nu}$ is

$$\theta = \log\{E(P_1)E(1-P_2)/E(P_2)E(1-P_1)\} - \nu,$$

whence

$$\theta = \log[-\{(1-\psi) + \log \psi\}/\{(1-\psi) + \psi \log \psi\}].$$

Two particular asymptotic values are

$$\theta = -\nu/3 + o(\nu) \quad \text{as } \nu \to 0,$$

and

$$\theta \sim -\log|\nu| \quad \text{as } |\nu| \to \infty.$$

Thus the bias due to collapsing the three-dimensional contingency table of pair, treatment, and response into a table of treatment and response only is over 30 percent when ν is near zero, but relatively small for large values of $|\nu|$.

9 MULTIVARIATE DATA

9.1 Multidimensional contingency tables

When the units of a population are classified by several methods applied simultaneously, the groups which result can be represented by the cells of a generalized rectangular lattice, where the axes refer to the methods of classification. The corresponding array of cell frequencies is known as a multidimensional contingency table. A table of d dimensions may have $1, 2, \ldots, d$ axes which correspond to responses, and the remainder then correspond to factors. Thus there are d basic types of table. An algorithm for printing multidimensional tables is given by Haberman (1973a).

Example 9.1 The data in Table 9.1 are taken from a report by Kihlberg, Narragon and Campbell (1964) which studied the relationship between car

TABLE 9.1 *Accidents to solitary drivers in two weights of car classified by accident type, accident severity, and driver ejected*

Car weight	Driver ejected	Accident classification			
		Collision		Rollover	
		Not severe	Severe	Not severe	Severe
Small	No	350	150	60	112
	Yes	26	23	19	80
Standard	No	1878	1022	148	404
	Yes	111	161	22	265

size and accident injuries. We follow Bhapkar and Koch (1968) in considering only drivers who were alone, and in condensing the original classifications of car weight, accident type, and accident severity to two categories each. The report presents a lengthy and detailed discussion of the statistical material. An outline analysis based only on Table 9.1 is as follows. There is no information about the numbers of small and standard cars at risk, and the totals of 820 and 4011 respectively may not be representative. Car weight is therefore taken as a factor, and accident type, accident severity, and driver ejected as responses. However, accident type affects accident severity, and both affect driver ejected.

Example 9.2 Minnesota high-school graduates of June 1938 were classified with respect to (i) position by thirds in high-school graduating class, (ii) post high-school status in April 1939, (iii) sex, and (iv) father's occupational level in seven categories. The categories of post high-school status are (C) enrolled in college, (N) enrolled in non-collegiate school, (E) employed full-time, and (O) other. Table 9.2 presents the data, which are quoted from

TABLE 9.2 *Minnesota high-school graduates classified by four methods*

Sex	Father's occupational level	High-school rank											
		Lowest third				Middle third				Upper third			
		C	N	E	O	C	N	E	O	C	N	E	O
Male	1	87	3	17	105	216	4	14	118	256	2	10	53
	2	72	6	18	209	159	14	28	227	176	8	22	95
	3	52	17	14	541	119	13	44	578	119	10	33	257
	4	88	9	14	328	158	15	36	304	144	12	20	115
	5	32	1	12	124	43	5	7	119	42	2	7	56
	6	14	2	5	148	24	6	15	131	24	2	4	61
	7	20	3	4	109	41	5	13	88	32	2	4	41
Female	1	53	7	13	76	163	30	28	118	309	17	38	89
	2	36	16	11	111	116	41	53	214	225	49	68	210
	3	52	28	49	521	162	64	129	708	243	79	184	448
	4	48	18	29	191	130	47	62	305	237	57	63	219
	5	12	5	10	101	35	11	37	152	72	20	21	95
	6	9	1	15	130	19	13	22	174	42	10	19	105
	7	3	1	6	88	25	9	15	158	36	14	19	93

Hoyt, Krishnaiah and Torrance (1959). Their objective was to study the dependence of post high-school status on the other variables. An analysis is also given by Ku and Kullback (1968), who point out that there is an unusually large number of girls as compared with boys for the third level of father's occupation.

9.2 Principles of analysis

The methods of analysis for a three-dimensional table can be extended without difficulty to multidimensional tables. They need to be supplemented by procedures which distinguish between the large number of statistical models available in four or more dimensions. Thus the problems are essentially the same as those which arise in the analysis of continuous measurements when there are many explanatory variables or factors.

We begin by changing the notation of section 7.2 in order to facilitate the generalizations that are required. Consider a three-dimensional table, and define sets of parameters as follows:

$$\theta = \log p_{rst}, \quad \theta_a^R = \log(p_{ast}/p_{rst}), \quad \theta_{ab}^{RS} = \log(p_{abt}p_{rst}/p_{ast}p_{rbt}),$$

and

$$\theta_{abc}^{RST} = \log(p_{abc}p_{ast}p_{rbt}p_{rsc}/p_{abt}p_{asc}p_{rbc}p_{rst}).$$

The sets of parameters $\{\theta_b^S\}$, $\{\theta_c^T\}$, $\{\theta_{ac}^{RT}\}$, and $\{\theta_{bc}^{ST}\}$ are similarly defined. Suppose that consideration is restricted to *hierarchical* models, which are such that the following property holds: if a set of parameters is zero, then any other set is zero for which the indices contain those of the original set. Thus

$$\theta_a^R = 0 \quad \text{implies} \quad \theta_{ab}^{RS} = \theta_{ac}^{RT} = \theta_{abc}^{RST} = 0,$$

$$\theta_{ab}^{RS} = 0 \quad \text{implies} \quad \theta_{abc}^{RST} = 0,$$

and similarly for other combinations of indices. Every model can be fitted to the observed frequencies by fixing a set of marginal totals and using the iterative scaling procedure on a table with unit frequencies. The relationship between the parameters which are zero and the marginal totals which are fixed is shown in Table 9.3 for combinations of parameters which typify

TABLE 9.3 *Hierarchical models for three-way classifications*

Parameters equal to zero	Number of parameters	Marginal totals fixed
$\{\theta_{abc}^{RST}\}$	$(r-1)(s-1)(t-1)$	$\{N_{ojk}\}, \{N_{iok}\}, \{N_{ijo}\}$
$\{\theta_{abc}^{RST}\}, \{\theta_{bc}^{ST}\}$	$r(s-1)(t-1)$	$\{N_{iok}\}, \{N_{ijo}\}$
$\{\theta_{abc}^{RST}\}, \{\theta_{bc}^{ST}\}, \{\theta_{ac}^{RT}\}$	$(rs-1)(t-1)$	$\{N_{ijo}\}, \{N_{ook}\}$
$\{\theta_{abc}^{RST}\}, \{\theta_{bc}^{ST}\}, \{\theta_{ac}^{RT}\}, \{\theta_c^T\}$	$rs(t-1)$	$\{N_{ijo}\}$
Not θ, $\{\theta_a^R\}, \{\theta_b^S\}, \{\theta_c^T\}$	$rst-r-s-t+2$	$\{N_{ioo}\}, \{N_{ojo}\}, \{N_{ook}\}$
Not θ, $\{\theta_a^R\}, \{\theta_b^S\}$	$rst-r-s+1$	$\{N_{ioo}\}, \{N_{ojo}\}$
Not θ, $\{\theta_a^R\}$	$r(st-1)$	$\{N_{ioo}\}$
Not θ	$rst-1$	N

all the possibilities in a three-way classification. Under conditions which are related to the effects of collapsing multidimensional contingency tables, maximum-likelihood estimates of the expected frequencies can be expressed in terms of the estimates for tables of lower dimensionality (Goodman, 1970; Haberman, 1970, ch. V: Bishop, 1971).

We now turn to the selection of an appropriate hierarchical model. The parameters of a saturated model are estimated by substituting $N_{ijk} + \frac{1}{2}$ for p_{ijk} in the expressions given above. For reasons which are explained in sections 5.2 and 7.5, the variances of the resulting estimators are themselves estimated by treating the variates $\{\log(N_{ijk} + \frac{1}{2})\}$ as if they were mutually independent and such that

$$\text{var}\{\log(N_{ijk} + \tfrac{1}{2})\} = 1/(n_{ijk} + \tfrac{1}{2}).$$

In this way we obtain standardized contrasts of the form $\hat{\theta}^{ST}_{bc} / \{\text{var}(\hat{\theta}^{ST}_{bc})\}^{\frac{1}{2}}$. Each is approximately $\mathfrak{N}(0, 1)$ if all the parameters are zero. Although the standardized contrasts are not independent, we can identify those which are significantly large by ranking them in order of increasing absolute value and plotting them against the quantiles of a half-normal distribution. If all the parameters were zero then the points would lie close to a line with unit slope, so that departures from the line indicate parameters which should be considered for inclusion in the model. The technique of half-normal plotting was introduced by Daniel (1959) in the context of factorial experiments, and first applied to contingency tables by Cox and Lauh (1967).

Another method of selecting an appropriate hierarchical model is as follows. Suppose that $\boldsymbol{\phi}_1$ and $\boldsymbol{\phi}_2$ are vectors with elements from the parameters θ, $\{\theta^R_a\}, \ldots, \{\theta^{RST}_{abc}\}$. A statistic for testing $\boldsymbol{\phi}_2 = \mathbf{0}$ given $\boldsymbol{\phi}_1 = \mathbf{0}$ against $\boldsymbol{\phi}_2 \neq \mathbf{0}$ given $\boldsymbol{\phi}_1 = \mathbf{0}$ is

$$Y^2 = 2\,\Sigma\, m_1 \log(m_1/m_{12}),$$

where m_1 and m_{12} are estimators of expected frequencies under $\boldsymbol{\phi}_1 = \mathbf{0}$ and $\boldsymbol{\phi}_1 = \mathbf{0}, \boldsymbol{\phi}_2 = \mathbf{0}$ respectively. The number of degrees of freedom for the approximating χ^2-distribution is the number of parameters in $\boldsymbol{\phi}_2$. We can therefore begin with a model which has been provisionally accepted and consider another which involves fewer parameters, a procedure of backward elimination. Alternatively, we can move to a model which involves more parameters, a procedure of forward selection, by calculating which of the corresponding hypotheses provides the best fit to the data. Stepwise procedures of this kind originated in the context of regression analysis, and are described by Draper and Smith (1966). They were first applied to contingency tables by Goodman (1971).

Deviations from any proposed model can be detected by an analysis of standardized residuals, which are typified by

$$e_{ijk} = (N_{ijk} - m_{ijk})/m^{\frac{1}{2}}_{ijk},$$

where m_{ijk} is an expected frequency under the model. The asymptotic distribution of $\{e_{ijk}\}$ is discussed by Haberman (1973b), and this work leads to the definition of adjusted residuals which are approximately $\mathfrak{N}(0, 1)$. We can identify the adjusted residuals which are significantly large by plotting the ordered values against the quantiles of a standard normal distribution. Further details are given by Cox and Snell (1968) and Haberman (1973c).

When a model has been finally accepted, the parameters can be estimated by substituting m_{ijk} for p_{ijk} in the expressions given above. The variances of the resulting estimators are themselves estimated by treating the variates $\{\log m_{ijk}\}$ as if they were mutually independent and such that

$$\text{var}\,\{\log m_{ijk}\} = 1/m_{ijk}.$$

9.3 Multivariate binary data

There are many experimental situations where each response or factor is defined by a classification with two categories. Methods for the analysis of multivariate binary data therefore merit special attention. When the general approach outlined in the previous section is specialized for binary data, another change of notation is convenient. We proceed to give two illustrations.

Suppose first that each of d binary classifications is a response. We assume that the frequencies $\{N_{ijk\,\ldots\,w}\}$ of a contingency table of order 2^d have a multinomial distribution with index n and probabilities $\{p_{ijk\,\ldots\,w}\}$ respectively, where $i, j, k, \ldots, w = 1, 2$. Define parameters as follows:

$$\theta_r = \log(p_{2\ldots212\ldots2}/p_{222\ldots2}),$$

where the suffixes are all 2 except that the probability in the numerator has a 1 in the rth position ($r = 1, 2, \ldots, d$). Next,

$$\theta_{rs} = \log(p_{2\ldots212\ldots212\ldots2}\,p_{222\ldots2}/p_{2\ldots212\ldots222\ldots2}\,p_{2\ldots222\ldots212\ldots2}),$$

where the suffixes are all 2 except that the probability in the numerator has 1 in both the rth and sth positions, while the probabilities in the denominator have 1 in the rth and sth positions respectively ($s = 2, 3, \ldots, d;\, r < s$). Other parameters $\theta_{rst}, \ldots, \theta_{12\ldots d}$ are defined in the same way. The distribution of $\{N_{ijk\ldots w}\}$ is

$$\begin{aligned}&(n!\,p_{222\ldots2}^{n}/\Pi_{ijk\ldots w}\,n_{ijk\ldots w}!)\\ &\quad\times\exp(n_{100\ldots0}\theta_1 + n_{010\ldots0}\theta_2 + \ldots + n_{00\ldots01}\theta_d\\ &\quad+ n_{110\ldots0}\theta_{12} + n_{1010\ldots0}\theta_{13} + \ldots + n_{00\ldots011}\theta_{d-1,d}\\ &\quad+ \ldots + n_{111\ldots1}\theta_{123\ldots d}).\end{aligned}$$

Distributions and procedures for problems of inference concerning $\{\theta_r\}$, $\{\theta_{rs}\}, \ldots$ can now be derived as before.

Secondly, we consider a set of binary responses indexed by the treatment combinations of a 2^{d-1} factorial experiment. This case is discussed by Zelen (1972) using an operational calculus. He obtains exact procedures of inference and approximations to them. An alternative approach is to express the distribution of $\{N_{ijk\ldots w}\}$ in the form

$$\Pi_{ijk\ldots w}(n_{0jk\ldots w}!\, p_{2jk\ldots w}{}^{n_{0jk\ldots w}}/n_{ijk\ldots w}!)$$
$$\times \exp(n_{100\ldots 0}\theta_1 + n_{110\ldots 0}\theta_{12} + \ldots + n_{10\ldots 01}\theta_{1d}$$
$$+ n_{1110\ldots 0}\theta_{123} + \ldots + n_{111\ldots 1}\theta_{123\ldots d}).$$

We interpret successive parameters here as a logistic transform when all factors are at level 2, the main effects on this transform of factors 2, 3, . . . , d, the interaction effect of factors 2 and 3, . . . , and the interaction effect of factors 2, 3, . . . d, respectively.

The analysis of multivariate binary data is much simplified if the parameters decrease in magnitude as the number of suffixes increases. When this situation does not obtain, matters may be improved by changing the definition of some responses, a technique suggested by Goodman (1971). An example due to Cox (1972) is as follows. Consider a sample of pairs consisting of husband and wife, and define

$$R = \begin{cases} 1 & \text{if husband votes Labour} \\ 0 & \text{otherwise} \end{cases}, \quad S = \begin{cases} 1 & \text{if wife votes Labour} \\ 0 & \text{otherwise.} \end{cases}$$

There are 2^2 possibilities. Another way of representing them is to define

$$R' = \begin{cases} 1 & \text{if husband and wife discordant} \\ 0 & \text{otherwise} \end{cases}, \quad S' = R.$$

A third way is to take $R'' = R'$, $S'' = S$. All other transformations of (R,S) can be obtained from these three pairs. When there are d binary responses, the choice of a transformation is equivalent to the choice of a set of generators for an Abelian group of order 2^d. For such transformations, Bloomfield (1972) shows that a log-linear model for the original data becomes another log-linear model with the coefficients permuted. He gives a modified version of the iterative scaling procedure to take account of the fact that a hierarchical model for the transformed data does not in general correspond to a hierarchical model for the original data. This line of approach can lead to a succinct description of the data, but one which may be difficult to interpret in practical terms.

9.4 Allocation rules

Suppose that a multivariate response $(R, S, \ldots)$ can be observed in each of two populations Ω_1 and Ω_2, and denote a typical observation $(i, j, \ldots)$ by $\mathbf{w}$. Write

$$p(R = i, S = j, \ldots) = \begin{cases} f_1(\mathbf{w}) & \text{in } \Omega_1, \\ f_2(\mathbf{w}) & \text{in } \Omega_2. \end{cases}$$

The problem is to find a rule δ for allocating an observation $\mathbf{w}$ to either Ω_1 or Ω_2. When the probability distributions $f_1(.)$ and $f_2(.)$ are exactly known, we proceed as follows. Let the prior probabilities that $\mathbf{w}$ comes from Ω_1 or Ω_2 be π_1 and π_2 respectively. Denote by $\alpha_u(\delta)$ the probability that rule δ misallocates an item from Ω_u chosen at random ($u = 1, 2$). Then the probability of misallocation, say $\rho(\delta)$, is given by

$$\rho(\delta) = \pi_1\alpha_1(\delta) + \pi_2\alpha_2(\delta) .$$

The posterior probabilities that $\mathbf{w}$ comes from Ω_1 or Ω_2 can be obtained from Bayes's theorem, whence we find that

$$\log\{p(\Omega_1 \mid \mathbf{w})/p(\Omega_2 \mid \mathbf{w})\} = \log(\pi_1/\pi_2) + \log\{f_1(\mathbf{w})/f_2(\mathbf{w})\}.$$

This result can be expressed in words as

log posterior odds (on Ω_1 versus Ω_2)

= log prior odds + log likelihood ratio.

A natural allocation rule is to assign $\mathbf{w}$ to Ω_1 or Ω_2 according as the log posterior odds is positive or negative, respectively. This rule, say ζ, minimizes the probability of misallocation (T.W. Anderson, 1958, sect. 6.3). When $\pi_1 = \pi_2 = \frac{1}{2}$, ζ allocates $\mathbf{w}$ to Ω_1 or Ω_2 according as the log likelihood ratio is positive or negative, respectively.

Since the probability distributions $f_1(.)$ and $f_2(.)$ are unknown, the allocation rule must be constructed on the basis of samples taken from the two populations. A change of notation is convenient. Denote the possible observations by $w_1, w_2, \ldots, w_t$. A sample of size n_u is taken from Ω_u and the frequency of w_k is N_{uk}, where $u = 1, 2$, and $k = 1, 2, \ldots, t$. The frequencies $\{N_{uk}\}$ have a multinomial distribution with index n_u and probabilities $\{p_{uk}\}$ respectively.

Suppose that t is small, and nothing is assumed about the structure of $\{p_{uk}\}$. The estimated value of p_{uk} is $\hat{p}_{uk} = n_{uk}/n_u$, and the estimated likelihood ratio is $\hat{p}_{1k}/\hat{p}_{2k}$. Denote by ζ^* the corresponding allocation rule. Then ζ^* has the following properties (Cochran and Hopkins, 1961; Hills, 1966). First, the expected value of the actual probability of misclassification, i.e. $E\,\{\rho(\zeta^*)\}$, exceeds the theoretical optimum probability $\rho(\zeta)$. Secondly, the expected value of the estimated probability of misclassification, i.e. $E\,\{\hat{\rho}(\zeta^*)\}$, is less than the theoretical optimum probability $\rho(\zeta)$.

When t is large, and nothing is assumed about the structure of $\{p_{uk}\}$, various modifications of the likelihood-ratio rule have been proposed and examined (Hills, 1967). For example, suppose that $f_1(.)$ and $f_2(.)$ refer to multivariate binary data. A nearest neighbour rule of order 1 is then based not only on the frequencies with which w_k is observed in each sample, but also on the frequencies for categories which differ from the kth in respect of just one of the binary responses.

The structure of $\{p_{uk}\}$ can be taken into account by assuming that each system of probabilities is represented by a log-linear model. This enables us to express the log posterior odds as a linear combination of unknown parameters. The estimation of these parameters by maximum likelihood on the basis of samples taken from each population separately is a problem discussed by J.A. Anderson (1972).

9.5 Numerical example

Example 9.3 The data in Table 9.1 are described at the beginning of this chapter. A preliminary analysis is made by estimating the main effects and interaction of ejection and severity for each combination of weight and type.

	Driver ejected $\log(n_{22}/n_{12})$		Accident severity $\log(n_{12}/n_{11})$		Interaction $\log(n_{11}n_{22}/n_{12}n_{21})$	
	Coll.	Roll.	Coll.	Roll.	Coll.	Roll.
Small	−1·88	−0·34	−0·85	0·62	0·72	0·81
Standard	−1·85	−0·42	−0·61	1·00	0·98	1·48

The following tentative conclusions emerge. Ejections occur at the same rate for each weight of car, but are much more likely in rollovers. Accidents tend to be somewhat more severe with standard cars and much more severe in rollovers. Ejections increase with severity, particularly in accidents which involve rollovers of standard cars.

A formal analysis begins by testing the hypothesis that weight has no effect on the interaction of type, severity, and ejection. The expected frequencies are as follows. They are calculated by fixing the 3D marginal totals and using ISP, beginning with a table in which all frequencies are 1.

348·392	151·608	61·608	110·392
27·608	21·392	17·392	81·608
1879·608	1020·392	146·392	405·608
109·392	162·608	23·608	263·392

Since $Y^2 = 0{\cdot}670$ (1 d.f.), the 4D interaction is taken as zero. Consider next the interaction type × severity × ejection, which now has the same value in the 3D tables for each weight. The expected frequencies for zero $T \times S \times E$ are obtained by applying ISP to each of the 3D tables with initial frequencies all 1 and fixed 2D totals. A comparison with the expected frequencies for zero 4D interaction gives $Y^2 = 2{\cdot}940$ (1 d.f.). The other 3D interactions are tested in the same way. None differ significantly from zero, but some variation in $S \times E$ is confirmed.

Interaction		Y^2
$W \times T \times S \times E$		0·670
$W \times S \times E$	2·924	
$T \times S \times E$	2·940	6·655
$W \times T \times E$	1·102	(see below)
$W \times T \times S$	1·222	

All $3D$ interactions are now taken as zero. The expected frequencies are as follows. They are calculated by fixing the $2D$ marginal totals and using ISP.

359·130	140·237	57·346	115·288
23·698	25·935	14·826	83·540
1875·040	1025·593	144·484	406·882
107·132	164·235	32·344	255·290

A comparison of these frequencies with the previous set gives $Y^2 = 6{\cdot}655$ (4 d.f.). At this stage there are few null hypotheses left that would not be rejected out of hand. We therefore turn to the calculation of estimates and standard errors for $2D$ effects. For example, severity × ejection is estimated as

$$\log\{(359{\cdot}130)\,(25{\cdot}935)/(140{\cdot}237)\,(23{\cdot}698)\},$$

and the estimated variance is the reciprocal of

$$(359{\cdot}130^{-1} + 25{\cdot}935^{-1} + 140{\cdot}237^{-1} + 23{\cdot}698^{-1})^{-1} + \text{three similar terms.}$$

Interaction	$S \times E$	$W \times T$	$W \times E$	$W \times S$	$T \times E$	$T \times S$
Estimate	1·031	−0·729	−0·144	0·337	1·366	1·638
Standard error	0·099	0·095	0·110	0·086	0·092	0·083

Much of the preliminary analysis is now confirmed and numerical values have been attached to the relevant parameters.

EXERCISES

1 N has a Poisson distribution with parameter μ. Take $\log p\,(N=j)$, use Stirling's formula

$$\log j! \sim (j+\tfrac{1}{2})\log j - j + \tfrac{1}{2}\log(2\pi),$$

substitute $j = \mu + x\mu^{\frac{1}{2}}$, and expand in powers of x. Hence show that

$$p\,(N=j) \sim (2\pi\mu)^{-\frac{1}{2}} \exp\{-(j-\mu)^2/2\mu\} + O\,(\mu^{-1}).$$

2 N has a Poisson distribution with parameter μ. Obtain an expansion for var $\{(N+c)^{\frac{1}{2}}\}$ in powers of μ^{-1}, and show that the coefficient of μ^{-1} is zero when $c = \frac{3}{8}$. Find a similar expansion for var $\{N^{\frac{1}{2}} + (N+1)^{\frac{1}{2}}\}$.

3 A sample of size r is taken from $\mathcal{P}(\mu)$, and the sample mean is M. Define

$$b\,(M;r,j) = \begin{cases} \binom{rM}{j}\left(\frac{1}{r}\right)^j\left(1-\frac{1}{r}\right)^{rM-j} & \text{for } rM \geqslant j\,, \\ 0 & \text{otherwise}\,. \end{cases}$$

Show that $b\,(M;r,j)$ is the unique minimum-variance unbiased estimator of $p(N=j)$.

(Barton, 1961; Glasser, 1962; Patil, 1963)

4 The independent variates N_1 and N_2 are $\mathcal{P}(\mu_1)$ and $\mathcal{P}(\mu_2)$ respectively. Denote μ_2/μ_1 by ρ. Conditional on $N_1 + N_2 = n$, show that $(N_1\rho - N_2)/(n\rho^{\frac{1}{2}})$ converges to $\mathcal{N}(0,1)$ as $n \to \infty$. Hence obtain a confidence interval for ρ with confidence coefficient approximately $1-\epsilon$.

5 A sample of size n_i is taken from $\mathcal{P}(\mu_i)$ and the sample mean is M_i $(i = 1,2)$. Show that the variance of $M_1 - M_2$ may be estimated by $M_1/n_2 + M_2/n_1$ if $\mu_1 = \mu_2$ and by $M_1/n_1 + M_2/n_2$ if $\mu_1 \neq \mu_2$.

(Kalbfleisch and Sprott, 1973)

6 A sample of size r is taken from $\mathcal{P}(\mu)$, and F_j is the frequency of the event $N = j$. Show that the joint distribution of $F_0, F_1, F_2, \ldots,$ conditional on $rM = n$, is given by

$$\frac{n!}{r^n} \frac{r!}{f_0!f_1!f_2!\ldots} \frac{1}{(2!)^{f_2}(3!)^{f_3}\ldots}$$

(Fisher, 1950)

7 A vector variate **y** has n elements which are mutually independent and have unit variance. Define $\mathbf{x} = \mathbf{B}\mathbf{y}$, where $\mathbf{B}$ is a matrix of order $m \times n$ and rank n. Thus the dispersion matrix of **x** is $\mathbf{B}\mathbf{B}^T$, which is singular if $m > n$. According to a theorem of Bose given by Pringle and Rayner (1971, p.25), the matrix $\mathbf{B}^T(\mathbf{B}\mathbf{B}^T)^{g_1}\mathbf{B}$ is unique, idempotent, symmetric, and has the same rank as **B**. Deduce that

$$\mathbf{y}^T\mathbf{y} = \mathbf{x}^T(\mathbf{B}\mathbf{B}^T)^{g_1}\mathbf{x}.$$

Extend this result to the case where **y** has a non-singular dispersion matrix **K**.

(D.A. Evans, personal communication)

8 Suppose that $N_1, N_2, \ldots, N_r$ are mutually independent Poisson variates with parameters $\mu_1, \mu_2, \ldots, \mu_r$, respectively. Let $\phi(.)$ be a differentiable function such that

$$\{\phi(\mu + x\mu^{\frac{1}{2}}) - \phi(\mu)\}/\mu^{\frac{1}{2}}\phi'(\mu) = x + O(\mu^{-\frac{1}{2}})$$

for every fixed x as $\mu \to \infty$. Show that the variate

$$\{\phi(N_i) - \phi(\mu_i)\}/\mu_i^{\frac{1}{2}}\phi'(\mu_i)$$

converges in distribution to $\mathcal{N}(0,1)$ as $\mu_i \to \infty$. Deduce that the kernel of the log likelihood function, in the limit as $\mu_i \to \infty$ for all i, is asymptotically equal to $-\frac{1}{2}W$, where

$$W = \Sigma\{\phi(N_i) - \phi(\mu_i)\}^2/N_i\{\phi'(N_i)\}^2.$$

9 The conditions are the same as in the previous exercise. Consider the hypothesis that $\mu_1, \mu_2, \ldots, \mu_r$ have a common value, say μ. Show that the minimum of W with respect to μ is distributed as χ^2_{r-1} in the limit as $\mu \to \infty$. Obtain the expressions for the minimum which correspond to

$$\text{(a) } \phi(\mu) = \mu, \quad \text{(b) } \phi(\mu) = \log \mu.$$

Verify that the corresponding values for the data of Table 2.1 are

(a) 20·1, (b) 21·0.

10 A random sample of size r is taken from $\mathcal{P}(\mu)$, and the observed values are $n_1, n_2, \ldots, n_r$. Suppose that μ has a prior distribution with density function

$$g(\mu) \propto e^{-a\mu}\,\mu^{b-1}$$

for $\mu \geqslant 0$, and otherwise zero. Show that the posterior distribution of μ belongs to the same family, and deduce that the posterior distribution of $2(a + r)\mu$ is $\chi^2_{2(b+rm)}$, where $rm = \Sigma\, n_i$. Hence obtain a Bayesian method of making inferences about the ratio of two Poisson means by using the F-distribution.

(Lindley, 1965, sect. 7.3)

11 The data in Table E.1 have been extracted from Matui (1932). They give, for each of eight rectangular areas measuring 1500 m by 1000 m,

TABLE E.1 *Number of houses per square of side 100 m in each of eight rectangular areas measuring 1500 m by 1000 m*

Area	Number of houses per square									
	0	1	2	3	4	5	6	7	8	9
1	78	55	13	3	1	–	–	–	–	–
2	77	53	18	2	–	–	–	–	–	–
3	76	44	23	5	1	1	–	–	–	–
4	75	47	22	4	1	1	–	–	–	–
5	69	45	24	9	2	1	–	–	–	–
6	78	47	18	5	1	1	–	–	–	–
7	64	54	26	3	1	–	–	1	–	1
8	67	53	24	4	2	–	–	–	–	–

the frequency distribution of the number of houses per square of side 100 m. Assuming that the distribution for each area is Poisson, but that different areas may have different parameters, test the hypothesis that the parameters of the areas have the same value.

12 The time of delivery for selected cases in Birmingham during 1950 and 1951 is studied by Charles (1953). Births are classified as institutional and domiciliary, and further divided by the period of the year and the hour of the day. The following frequencies are those for institutional births from 8 March to 4 April, and they correspond to the 24 hours beginning at 12 midnight. Test the hypothesis that the data form a sample from a Poisson distribution against the alternative that independent Poisson variates have a daily peak in mean value.

42, 54, 52, 65, 50, 44, 42, 42, 28, 36, 36, 32,

39, 42, 33, 31, 39, 28, 41, 37, 31, 35, 39, 43.

13 The variates $N_1, N_2, \ldots, N_r$ have a multinomial distribution with index n and probabilities $p_1, p_2, \ldots, p_r$. Simultaneous confidence intervals typified by $l_i < p_i < u_i$ are constructed as follows. Write k for the upper 100 α/r percent point of χ_1^2. Then l_i and u_i are the values of p_i which satisfy the equation

$$(n_i - np_i)^2 = knp_i(1 - p_i).$$

Prove that the intervals have the following property. As $n \to \infty$, the probability is at least $1 - \alpha$ that all the intervals include the corresponding true values of the parameters.

(Quesenberry and Hurst, 1964; Goodman, 1965).

14 A variate T has a binomial distribution with index n and probability θ. Show that

$$E\,[\log\{(T+\tfrac{1}{2})/(n-T+\tfrac{1}{2})\}] = \log\{\theta/(1-\theta)\} + O(n^{-2}),$$

and

$$E\,\{(T+\tfrac{1}{2})^{-1} + (n-T+\tfrac{1}{2})^{-1}\} = \text{var}\,[\log\{(T+\tfrac{1}{2})/(n-T+\tfrac{1}{2})\}] + O(n^{-3}).$$

(Haldane, 1955; Anscombe, 1956; Gart and Zweifel, 1967)

15 A variate T has a binomial distribution with index n and probability θ. Verify that

$$p(T \geqslant t) = \int_0^\theta x^{t-1}(1-x)^{n-t}\,dx/B(t, n-t+1),$$

and

$$p(T > t) = \int_0^\theta x^{t}(1-x)^{n-t-1}\,dx/B(t+1, n-t).$$

The result suggests that we treat θ as a beta variate with density function

$$h(x) = x^{t-\frac{1}{2}}(1-x)^{n-t-\frac{1}{2}}/B(t+\tfrac{1}{2}, n-t+\tfrac{1}{2}).$$

Prove that $\theta\{2(n-t)+1\}/(1-\theta)(2t+1)$ is then distributed as F with $2t+1$ and $2(n-t+1)$ degrees of freedom. Deduce that $\log\{\theta/(1-\theta)\}$ is approximately normal with mean $\log\{(t+\frac{1}{2})/(n-t+\frac{1}{2})\}$ and variance $(t+\frac{1}{2})^{-1} + (n-t+\frac{1}{2})^{-1}$.

Show that the same results are obtained from a Bayesian analysis which starts from the prior density function

$$g(\theta) = 1/\pi\{\theta(1-\theta)\}^{\frac{1}{2}}.$$

(Gart, 1966)

16 The units of a population are classified with respect to two binary responses. Let p_{ij} be the probability that a unit is placed in category i of the first response and category j of the second. Suppose that a measure of association between responses must satisfy the following two conditions :

(a) The measure is a function of p_{11}/p_{10} and p_{21}/p_{20}, or alternatively, of p_{11}/p_{01} and p_{12}/p_{02}.

(b) These alternative measures are equal.

Show that any such measure of association is a function of the cross-product ratio $p_{11}p_{22}/p_{12}p_{21}$.

(Edwards, 1963)

17 N_{11} has the distribution $h(n_{11}\,;n_{i0}, n_{0j}, \lambda)$. Show that

$$E\,(N_{11}N_{22} - e^{\lambda}N_{12}N_{21}) = 0,$$

and hence obtain

$$E\,(N_{11}\,;\lambda)\,E\,(N_{22}\,;\lambda) - e^{\lambda}E\,(N_{12}\,;\lambda)\,E\,(N_{21}\,;\lambda) = (e^{\lambda}-1)\,\mathrm{var}\,(N_{11}\,;\lambda).$$

(Birch, 1964)

18 Consider the distribution $h(n_{11}\,;n_{i0}, n_{0j}, \lambda)$ in the limit as $n_{10} \to \infty$ and $n \to \infty$, such that $n_{10}/n \to \alpha$. Denote $\alpha e^{\lambda}/(1-\alpha+\alpha e^{\lambda})$ by β. Prove that

$$\lim h(n_{11}\,;n_{i0}, n_{0j}, \lambda) = \binom{n_{01}}{n_{11}}\beta^{n_{11}}(1-\beta)^{n_{21}}.$$

(Harkness, 1965)

19 Consider the distribution $h(n_{11}\,;n_{i0}, n_{0j}, \lambda)$ in the limit as $n_{10} \to \infty$, $n_{01} \to \infty$ and $n \to \infty$, such that $n_{10}n_{01}/n \to \xi$. Denote ξe^{λ} by μ. Prove that

$$\lim h(n_{11}\,;n_{i0}, n_{0j}, \lambda) = \mu^{n_{11}}\,e^{-\mu}/n_{11}\,!$$

(Harkness, 1965)

20 The variates N_{11}, N_{12}, N_{21} and N_{22} are normally and independently distributed with variances m_{11}, m_{12}, m_{21} and m_{22} respectively. Construct the dispersion matrix of N_{11}, N_{10}, N_{01} and N. Use the properties of a normal multivariate distribution to show that the variance of N_{11}, conditional on the observed values of N_{10}, N_{01} and N, is

$$(m_{11}^{-1}+m_{12}^{-1}+m_{21}^{-1}+m_{22}^{-1})^{-1}.$$

21 N_{11} has the distribution $h(n_{11}\,;n_{i0}, n_{0j}, \lambda)$. Denote $\log\,(N_{11}N_{22}/N_{12}N_{21})$

by L, and write

$$V = (L-\lambda)(m_{11}^{-1}+m_{12}^{-1}+m_{21}^{-1}+m_{22}^{-1})^{-\frac{1}{2}}.$$

Under the conditions given in section 4.2, the variate

$$X = (N_{11}-m_{11})(m_{11}^{-1}+m_{12}^{-1}+m_{21}^{-1}+m_{22}^{-1})^{\frac{1}{2}}$$

converges in distribution to $\mathfrak{N}(0,1)$. Use this result to show that V converges to $\mathfrak{N}(0,1)$ under the same conditions.

22 The units of a sample are classified with respect to two binary responses. Define

$$\kappa = \log p_{22}, \quad \theta = \log(p_{12}/p_{22}), \quad \phi = \log(p_{21}/p_{22}),$$

and

$$\lambda = \log(p_{11}p_{22}/p_{12}p_{21}).$$

Thus the likelihood function is

$$n!\exp(n\kappa + n_{10}\theta + n_{01}\phi + n_{11}\lambda)/\Pi_{ij}\, n_{i}.$$

Maximize this function subject to

$$p_{11}+p_{12}+p_{21}+p_{22} = 1.$$

Hence show that the unconditional maximum-likelihood estimator of λ is

$$\hat{\lambda} = \log(N_{11}N_{22}/N_{12}N_{21}).$$

23 The conditions are the same as in the previous exercise. Show that the joint distribution of N_{10} and N_{01} is

$$n!\exp(n\kappa + n_{10}\theta + n_{01}\phi)\sum_{n_{11}=v}^{w}\exp(n_{11}\lambda)/\Pi_{ij}\, n_{ij}!$$

Prove that the maximum-likelihood estimator of λ from this distribution satisfies the equation

$$\lambda = \log\{E(N_{11};\lambda)E(N_{22};\lambda)/E(N_{12};\lambda)E(N_{21};\lambda)\}.$$

Verify that the equation is satisfied by $\lambda = 0$, and by all values of λ when $n_{10}\to\infty$, $n_{01}\to\infty$ and $n\to\infty$, such that $n_{10}/n\to\pi_{10}$ and $n_{01}/n\to\pi_{01}$.

24 Physicians with and without lung cancer are classified by smoking status in the table below, which is taken from Wynder and Cornfield (1953).

Smoking status	Lung cancer	Controls
Nonsmoker	3	11
Smoker	60	32

Find the maximum-likelihood estimate of λ from the conditional distribution, and obtain a confidence interval for λ with a confidence coefficient of at least 95 percent.

25 The units of a sample are classified with respect to two responses R and S. When the classifications are made correctly,

$$p(R=i, S=j) = p_{ij} \quad (i=1,2,\ldots,r;\ j=1,2,\ldots,s).$$

However, the data are subject to misclassification, which obtains independently for the two responses. The effect on R is that a unit truly in category i is recorded in category u with probability α_{iu}. There is a similar probability β_{jv} for S, and

$$\Sigma_u \alpha_{iu} = \Sigma_v \beta_{jv} = 1 \quad \text{for all } i \text{ and } j.$$

After misclassification,

$$p(R=u, S=v) = p'_{uv} \quad (u=1,2,\ldots,r;\ v=1,2,\ldots,s).$$

Show that

$$p'_{uv} = \Sigma_{ij}\, p_{ij}\alpha_{iu}\beta_{jv}.$$

26 The conditions are the same as in the previous exercise, and $r = s = 2$. Write

$$\lambda' = \log(p'_{11}p'_{22}/p'_{12}p'_{21}).$$

Prove that if $\lambda = 0$, then $\lambda' = 0$, but that otherwise $|\lambda'| < |\lambda|$. Deduce that the procedures for testing $\lambda \leqslant 0$ against $\lambda > 0$, or $\lambda = 0$ against $\lambda \neq 0$, remain valid under misclassification but that their powers are reduced.

(Bross, 1954; Newell, 1963; Walsh, 1963; Mote and Anderson, 1965; Assakul and Proctor, 1967)

27 Two treatments which each give rise to a binary response are compared as follows. The experiment requires n items, which are divided by a random partition into a group of n_{10} receiving the first treatment and a group of n_{20} receiving the second. Here n_{10} and n_{20} are fixed in advance. The frequencies obtained are shown in the table below:

	Success	Failure	Total
First treatment	N_{11}	N_{12}	n_{10}
Second treatment	N_{21}	N_{22}	n_{20}
Total	N_{01}	N_{02}	n

We require to test that the treatments are equally effective. Show that a valid test of this hypothesis is based on the distribution of N_{11} conditional on the observed value of N_{01}.

(Barnard, 1947; Pearson, 1947)

28 Treatments 1 and 2 are applied individually and together to groups of mice. The probabilities of death are p_1 for treatment 1 and p_2 for treatment 2. According to one hypothesis of joint action, the probability of death is $p_1 p_2$ when the treatments are applied together. Denote the corresponding sample sizes by n_{10}, n_{20}, n_{30} respectively. Let N_{i1} be the number of deaths in group i ($i = 1, 2, 3$), and N_{i2} the number of survivors. Show that the agreement of data and hypothesis can be tested by the statistic

$$X^2 = u^2 \left\{ \frac{N_{12}}{n_{10}(N_{11} + u)} + \frac{N_{22}}{n_{20}(N_{21} + u)} + \frac{N_{32}}{n_{30}(N_{31} - u)} \right\},$$

where u satisfies the quadratic equation

$$(n_{10} + u)(n_{20} + u)(N_{31} - u) = (N_{11} + u)(N_{21} + u)(n_{30} - u),$$

and is the root which lies between N_{31} and the negative value of the smaller of N_{11} and N_{21}.

(Fisher, appendix to Mather, 1940; Fisher, 1942)

29 A binary response is observed on individuals before treatment, and again after treatment. We require to test the hypothesis of equal proportional change, expressed by

$$p\,(1 \text{ after} \mid 2 \text{ before}) = p\,(2 \text{ after} \mid 1 \text{ before}).$$

Show that this hypothesis is identical with the hypothesis that U and V are independent, where U is response before treatment and V is change, classified as present or absent.

(B. Hajtman, personal communication)

30 Pairs of twins are classified with respect to the presence (A) or absence ($\bar{A}$) of a specified blood-group. The pairs ($A\bar{A}$) and ($\bar{A}A$) are indistinguishable. Denote the probabilities of ($\bar{A}\bar{A}$), ($\bar{A}A$), and (AA) pairs by θ_0, θ_1, and θ_2 respectively. According to the hypothesis of independence, these are the probabilities of a binomial distribution with index 2 and probability p_A. Deduce that $\theta_1^2 = 4\theta_0\theta_2$.

A sample of n pairs contains n_0, n_1, and n_2 in the categories ($\bar{A}\bar{A}$), ($\bar{A}A$), and (AA) respectively. The frequencies define an *intraclass* 2×2 table which can be represented in either of the forms shown below :

Pair table

	$\bar{A}$	A
$\bar{A}$	n_0	n_1
A	.	n_2

Component table

	$\bar{A}$	A
$\bar{A}$	$2n_0$	n_1
A	n_1	$2n_2$

We require to test the hypothesis of independence against the alternative of positive dependence, defined by $\theta_1^2 < 4\,\theta_0\theta_2$. Show that a uniformly most powerful test is based on the distribution of N_0 conditional on the observed frequencies of A and $\bar{A}$ individuals when independence holds, given by

$$n!\,(2n_0+n_1)!\,(n_1+2n_2)!\,2^{n_1}/(2n)!\,n_0!\,n_1!\,n_2!$$

(Armitage and Healy, 1957; Okamoto and Ishii, 1961; Altham, 1971b)

31 A 2 X 2 table with observed frequencies $\{n_{ij}\}$ is analysed by standard and Bayesian methods. In the standard analysis, define

$$\alpha = \sum_{x=n_{11}}^{w} h(x;n_{i0},n_{0j},0),$$

and

$$\beta = \sum_{x=n_{11}+1}^{w} h(x;n_{i0},n_{0j},0).$$

The Bayesian analysis is made using a prior distribution for $\{p_{ij}\}$ with density function

$$f(p_{11},p_{12},p_{21},p_{22}) \propto \Pi_{ij}\,p_{ij}^{t-1},$$

where $0 \leqslant t \leqslant 1$. Define

$$\pi = p\,(\Psi < 1 \mid N_{ij} = n_{ij}, \quad \text{all } i,j).$$

Show that $\beta < \pi < \alpha$. Deduce that the difference between the significance levels of the standard and Bayesian procedures is less than the probability of the data on the hypothesis $\lambda = 0$.

(Altham, 1969)

32 Suppose that $B_1, B_2, \ldots, B_s$ are mutually independent variates such that B_j has a binomial distribution with index t and probability p_j. Write $C = \Sigma\,B_j/s$. The problem is to test the hypothesis that

$$H: p_1 = p_2 = \ldots = p_s$$

against the alternative that the probabilities are not all equal. Show that the statistic X^2 is given by

$$X^2 = \Sigma\,(B_j - C)^2/C(1 - C/t).$$

When H is true, prove that X^2 converges in distribution to χ^2_{s-1} as $t \to \infty$. In this connection, X^2 is called the *binomial index of dispersion*.

33 The frequency distribution of the number of boys in 200 families of size 5 is given below.

Number of boys	0	1	2	3	4	5	Total
Frequency	5	26	65	64	31	9	200

Use the binomial index of dispersion to test the hypothesis that the probability of a boy is the same for each family.

34 The frequencies $\{N_{ij}\}$ form a contingency table of order $r \times s$. Define

$$X^2 = n\,\Sigma_{ij}(N_{ij} - n_{io}n_{oj}/n)^2/n_{io}n_{oj}\,.$$

Given that the distribution of $\{N_{ab}\}$ is $h\,\{n_{ab};\, n_{io}, n_{oj}, 0\}$, show that

$$E\,(X^2) = n(r-1)\,(s-1)/(n-1).$$

Introduce the following quantities:

$$\nu = (r-1)\,(n-r)/(n-1), \quad \mu = (s-1)\,(n-s)/(n-1),$$

$$\sigma = (n\,\Sigma\, n_{io}^{-1} - r^2)/(n-2), \quad \tau = (n\,\Sigma\, n_{oj}^{-1} - s^2)/(n-2).$$

Show that

$$\text{var}\,(X^2) = 2n(\nu - \sigma)\,(\mu - \tau)/(n-3) + n^2\sigma\tau/(n-1).$$

(Bartlett, 1937; Haldane, 1940; Dawson, 1954)

35 The distribution of a bivariate response (R, S) is defined by probabilities $\{p_{ij}\}$. Apply the iterative scaling procedure to a table with unit entries and marginal probabilities $\{p_{io}\}$ and $\{p_{oj}\}$. Show that the procedure leads to

$$p_{ij} = p_{io}\,p_{oj}$$

after rows and columns have each been scaled once.

36 The conditional distribution of $\{N_{ab}\}$ is given by $h\{n_{ab};\, n_{io}, n_{oj}, \lambda^0_{ab}\}$, where the parameters $\{\lambda^0_{ab}\}$ form a vector $\boldsymbol{\lambda}_0$. Write $\mathbf{V}$ for the (singular) dispersion matrix of $\{N_{ij}\}$, and $\mathbf{D}$ for a diagonal matrix with diagonal elements $\{m_{ij}(\boldsymbol{\lambda}_0)\}$, where these elements follow the same order as $\{N_{ij}\}$. Show that asymptotically

$$\mathbf{V}\mathbf{D}^{-1}\mathbf{V} = \mathbf{V}.$$

Deduce that $S_3 = X^2$ to the same order, where both statistics are those used to test the hypothesis $H : \boldsymbol{\lambda} = \boldsymbol{\lambda}_0$ against the alternative $K : \boldsymbol{\lambda} \neq \boldsymbol{\lambda}_0$.

37 The frequencies of a contingency table of order $r \times r$ have a multinomial distribution with index n and probabilities $\{p_{ij}\}$. We require to test the hypothesis of marginal homogeneity, according to which

$$p_{io} = p_{oi} \quad (i = 1, 2, \ldots, r).$$

Denote $N_{io} - N_{oi}$ by D_i. Show that

$$\text{var}(D_i) = n\{(p_{io} + p_{oi} - 2p_{ii}) - (p_{io} - p_{oi})^2\},$$

and

$$\text{cov}(D_i, D_j) = -n\{(p_{ij} + p_{ji}) + (p_{io} - p_{oi})(p_{jo} - p_{oj})\}.$$

Let $\mathbf{d}$ be the vector with elements $D_1, D_2, \ldots, D_{r-1}$, and the $\mathbf{V}$ the matrix with elements

$$v_{aa} = N_{ao} + N_{oa} - 2N_{aa} \qquad (a = 1, 2, \ldots, r-1).$$

$$v_{ab} = -(N_{ab} + N_{ba}) \qquad (a \neq b).$$

Define $\qquad X_S^2 = \mathbf{d}^T\mathbf{V}^{-1}\mathbf{d}.$

Prove that X_S^2 converges in distribution to χ^2_{r-1} as $n \to \infty$ when marginal homogeneity obtains.

(Stuart, 1955)

38 Another test of marginal homogeneity proposed by Bhapkar (1966) is based on the statistic

$$X_B^2 = \mathbf{d}^T(\mathbf{V} - \mathbf{d}\mathbf{d}^T/n)^{-1}\mathbf{d}.$$

Show that $\qquad X_B^2 = X_S^2/(1 - X_S^2/n).$

Deduce that the tests are asymptotically equivalent.

39 The data in Table 7.3 are described in section 7.1. Estimate the expected frequencies corresponding to the hypotheses of quasi-symmetry and marginal homogeneity. Test these hypotheses and comment on the results.

40 A sample of size n is taken from the distribution of a bivariate response (R, S). The observation on N_{11} is missing. Show that the hypothesis that R and S are independent can be tested as follows. Denote the existing totals for the first row, first column, and all frequencies by n'_{10}, n'_{01}, and n' respectively. Replace the missing frequency by

$$n'_{11} = n'_{10}n'_{01}/(n - n'_{10} - n'_{01}).$$

Calculate X^2 or Y^2 by the standard formula, treating n'_{11} as an observed frequency. Refer the test statistic to the distribution of $\chi^2_{(r-1)(s-1)-1}$.

(Watson, 1956)

41 An *upper triangular* contingency table is a table of order $r \times r$ in which cell (i, j) is empty when $i + j > r + 1$. Suppose that the probabilities for the non-empty cells satisfy the hypothesis of quasi-independence. Prove that the maximum-likelihood estimates of the expected frequencies are obtained by the following procedure:

$$m_{r1} = n_{r0}, \quad m_{1r} = n_{0r},$$

and

$$m_{st} = n_{s0}\, n_{0t}/(n - \sum_{i=s+1}^{r} n_{i0} - \sum_{j=t+1}^{r} n_{0j})$$

for any cell (s,t) on the diagonal where $s + t = r + 1$. Now subtract the diagonal estimates from the original marginal totals, and repeat the procedure with the resulting triangular table of order $(r - 1) \times (r - 1)$. Continue until all the expected frequencies have been estimated.

(Waite, 1915; Bishop and Fienberg, 1969)

42 A *lower triangular* contingency table is a table of order $r \times r$ in which cell (i,j) is empty when $i < j$. Show that the expected frequencies under quasi-independence are obtained as follows. Calculate

$$s_i = \sum_{j=1}^{i} (n_{0j} - n_{j0}), \quad x_i = n_{0(i+1)}/s_i, \quad \text{and} \quad y_{i+1} = \prod_{j=1}^{i} (1 + x_j) \text{ with } y_1 = 1\,.$$

Then the expected frequency for cell (i, j) is estimated by $\hat{\alpha}_i\hat{\beta}_j$, where

$$\hat{\alpha}_i = n_{i0}/y_i \quad \text{and} \quad \hat{\beta}_j = x_{j-1}y_{j-1} \quad (j = 2,3,\ldots,r) \text{ with } \hat{\beta}_1 = 1.$$

(Goodman, 1968)

43 The data in Table E.2 are quoted by Bishop and Fienberg (1969) and were collected from patients at the Massachusetts General Hospital by Jones and Poskanzer. On admission, patients were graded on a five-point scale increasing in severity from A to E according to physical disability following a stroke. They were graded on the same scale at the time of discharge. Since none of the patients had a second stroke, the grade on discharge could only be the same or better as on admission. Thus the frequencies are zero in all cells where the second state is worse than the first. Estimate the expected frequencies corresponding to the hypothesis of quasi-independence. Test this hypothesis, and interpret the results.

TABLE E.2 *Initial and final disability ratings of 121 stroke patients*

Initial state	Final state *A*	*B*	*C*	*D*	*E*	Total
E	11	23	12	15	8	69
D	9	10	4	1	0	24
C	6	4	4	0	0	14
B	4	5	0	0	0	9
A	5	0	0	0	0	5
Total	35	42	20	16	8	121

44 A sample of size n is taken from the distribution of a bivariate response (R, S). Suppose that $n_{11} + n_{12} = w$ is known, but not n_{11} and n_{12} separately. Under the hypothesis that R and S are independent, show that the maximum-likelihood estimates of the marginal probabilities are as follows:

$$\hat{p}_{io} = n_{io}/n \quad (i = 1, 2, \ldots, r), \qquad \hat{p}_{oj} = n_{oj}/n \quad (j = 3, 4, \ldots, s)$$

$$\hat{p}_{01} = (n_{01} + n_{02})(n_{01} - n_{11})/n\,(n_{01} + n_{02} - w).$$

$$\hat{p}_{02} = (n_{01} + n_{02})(n_{02} - n_{12})/n\,(n_{01} + n_{02} - w).$$

(Craig, 1953)

45 Each of n individuals is asked to rank each of r objects according to first, second, . . . , rth choice. Write p_{ij} for the probability that object i is assigned rank j, and let N_{ij} be the corresponding frequency. Then

$$p_{io} = 1 \quad \text{and} \quad N_{jo} = N_{oj} = n.$$

We require to test the hypothesis that

$$p_{ij} = 1/r \quad (i, j = 1,2, \ldots, r).$$

When this hypothesis holds, show that the statistic

$$(r-1)\,\Sigma_{ij}\,(N_{ij} - n/r)^2/n$$

converges in distribution to $\chi^2_{(r-1)^2}$ as $n \to \infty$.

(R.L. Anderson, 1959)

46 A classification with r categories is made in each of s populations. The probability is p_{ij} that a unit from the jth population is placed in the ith category, and the corresponding frequency is N_{ij}. We require to test the hypothesis

$$H : p_{ij} = \pi_{ij} \quad (i = 1, 2, \ldots, r;\ j = 1, 2, \ldots, s),$$

where the values $\{\pi_{ij}\}$ are specified. Consider the statistic

$$T^2 = \Sigma_i (N_{io} - e_i)^2 / e_i,$$

where

$$e_i = \Sigma_j n_{0j} \pi_{ij}.$$

If $n_{01}, n_{02}, \ldots, n_{0s}$ all $\to \infty$ in fixed proportions, show that the distribution of T^2 converges to $\Sigma_i \lambda_i Y_i^2$ when H is true, where $Y_1, Y_2, \ldots, Y_r$ are mutually independent $\mathfrak{N}(0, 1)$ variates and $\lambda_1, \lambda_2, \ldots, \lambda_r$ are the latent roots of the matrix with elements

$$\delta_{ih} - (e_i e_h)^{-\frac{1}{2}} \Sigma_j n_{0j} \pi_{ij} \pi_{hj} \quad (i,h = 1, 2, \ldots, r).$$

(Okamoto, 1963)

47 A sample of size n is taken from the distribution of a bivariate response (R, S). Suppose that the cell probabilities have a Dirichlet prior distribution with density function

$$f\{p_{ij}\} \propto \Pi_{ij} p_{ij}^{k\nu_{ij}-1}, \quad \text{where } \Sigma_{ij} \nu_{ij} = 1.$$

Denote by T_{ij} an estimator of the expected frequency np_{ij}, and let the loss function be $\Sigma_{ij} (T_{ij} - np_{ij})^2$. Show that the Bayes estimator of np_{ij} is

$$B_{ij} = n(n_{ij} + k\nu_{ij})/(n + k).$$

(Fienberg and Holland, 1970)

48 An insect passes successively through $(r - 1)$ stages during development and eventually reaches the rth stage. The durations of the first $(r - 1)$ stages are mutually independent non-negative random variables $T_1, T_2, \ldots, T_{r-1}$. Denote the time from hatching by t. Then the insect is in stage

$$\begin{aligned}
&1 \text{ if } T_1 > t,\\
&i \text{ if } T_1 + T_2 + \ldots + T_{i-1} \leqslant t\\
&\quad \text{and } T_1 + T_2 + \ldots + T_i > t \quad (i = 2, 3, \ldots, r - 1),\\
&r \text{ if } T_1 + T_2 + \ldots + T_{r-1} \leqslant t.
\end{aligned}$$

Suppose that T_i is $\mathfrak{N}(\mu_i, \sigma^2 \mu_i)$ where σ is small $(i = 1, 2, \ldots, r - 1)$. Derive the maximum-likelihood equations and the information matrix for the parameters $\mu_1, \mu_2, \ldots, \mu_{r-1}$ and σ.

(Aitchison and Silvey, 1957)

49 A binary response R is observed at each level of a factor $\mathcal{F}$. Suppose that

$$\operatorname{logit} p_{1j} = \alpha + \beta x_j \quad (j = 1, 2, \ldots, s).$$

Show that the distribution of $\Sigma N_{1j}x_j$ conditional on the observed value of N_{10} provides a uniformly most powerful unbiased test of the hypothesis $\beta = 0$ against the alternative $\beta \neq 0$. When $\beta = 0$, prove that

$$E(\Sigma_j N_{1j} x_j) = n_{10}\mu_x,$$

and

$$\operatorname{var}(\Sigma_j N_{1j} x_j) = n_{10}n_{20}\sigma_x^2/(n-1),$$

where $\mu_x = \Sigma_j n_{0j} x_j/n$ and $\sigma_x^2 = \Sigma_j n_{0j}(x_j - \mu_x)^2/n$.

50 Assume that the model of the previous exercise applies to the data in Table 3.3, with $x_j = -1, 0, 1$ for the three levels of tonsil size. Use a normal approximation to the distribution of $\Sigma_j N_{1j} x_j$ to test that $\beta = 0$. Compare the result with that obtained from X^2, and explain the difference.

51 A binary response R is observed at each level of a factor $\mathcal{F}$. Suppose that

$$p_{1j} = \alpha + \beta x_j \quad (j = 1, 2, \ldots, s).$$

Show that an unbiased estimator of β is

$$\beta^* = (n\, \Sigma_j N_{1j}x_j - N_{10}\, \Sigma_j n_{0j}x_j)/n^2\sigma_x^2,$$

and that an unbiased estimator of var (β^*) when $\beta = 0$ is

$$\operatorname{var}^*(\beta^*) = N_{10}N_{20}/n^2(n-1)\,\sigma_x^2,$$

where σ_x^2 is defined in Exercise 49. Derive a procedure for testing the hypothesis $\beta = 0$ against the alternative $\beta \neq 0$ for small values of β.

(Cochran, 1954; Armitage, 1955)

52 A bivariate response (R, S) is observed, where each response is ordered. Introduce scores $\{c_i\}$ for R and $\{d_j\}$ for S, and make the assumption that

$$\lambda_{ab} = \beta(c_a - c_r)(d_b - d_s) \quad (a = 1, 2, \ldots, r-1; b = 1, 2, \ldots, s-1).$$

Prove the following results:

(i) The conditional maximum-likelihood estimator of β satisfies the equation

$$\Sigma_{ij} N_{ij}c_i d_j = E(\Sigma_{ij} N_{ij} c_i d_j; \tilde{\beta}).$$

(ii) A uniformly most powerful unbiased test of the hypothesis $\beta = 0$ against the alternative $\beta \neq 0$ is obtained by referring $\Sigma_{ij} N_{ij} c_i d_j$ to its distribution conditional on the observed values of $\{N_{io}\}$ and $\{N_{oj}\}$.

(iii) When $\beta = 0$, the mean and variance of the conditional distribution are

$$E(\Sigma_{ij} N_{ij} c_i d_j; 0) = n\mu_c \mu_d,$$

and

$$\text{var}(\Sigma_{ij} N_{ij} c_i d_j; 0) = n\sigma_c^2 \sigma_d^2/(n-1),$$

where μ_c, μ_d, σ_c^2 and σ_d^2 are defined by analogy with μ_x and σ_x^2 in Exercise 49.

(iv) If β is small, then approximately

$$\tilde{\beta} = (n-1)\, \Sigma_{ij} N_{ij} (c_i - \mu_c)(d_j - \mu_d)/\Sigma_i n_{io}(c_i - \mu_c)^2\, \Sigma_j n_{oj}(d_j - \mu_d)^2,$$

and

$$\text{var}(\tilde{\beta}) = \{\Sigma_i n_{io}(c_i - \mu_c)^2\, \Sigma_j n_{oj}(d_j - \mu_d)^2/(n-1)\}^{-1}.$$

(Yates, 1948; Mantel, 1963; Birch, 1965)

53 Apply the model of the previous exercise to the data quoted in Table 3.4. The system of cross-product ratios corresponding to the scores suggested in section 6.1 is defined by the following table:

$e^{2\beta}$	e^{β}	1	$e^{-\beta}$	$e^{-2\beta}$
1	1	1	1	1
$e^{-2\beta}$	$e^{-\beta}$	1	e^{β}	$e^{2\beta}$

For values of β near zero, use the iterative scaling procedure to calculate expected frequencies. Hence obtain the following results.

(i) When $\beta = 0$, $Y^2 = 9{\cdot}037$ (8 d.f.)

(ii) The minimum value of Y^2 is 6·706 (7 d.f). This value is attained when $\beta = 0{\cdot}053$.

(iii) A confidence interval for β with confidence coefficient 95 percent is $(-0{\cdot}0151,\ 0{\cdot}1215)$. State your conclusions concerning the improvement in teacher's rating with improved homework conditions.

54 A binary response is observed at each level of a quantitative factor $\mathfrak{F}$. Suppose that the frequencies $\{N_{ij}\}$ are mutually independent Poisson variates with parameters $\{\mu_{ij}\}$ respectively, where

$$\mu_{1j} = \mu_{2j} + \theta + \beta(x_j - x_s).$$

Apply the minimum modified chi-square method to estimate $\{\mu_{2j}\}$, θ and β. Show that the estimators of θ and β are the same as those obtained by the minimum logit chi-square method.

55 Patients with a condition requiring treatment are classified by sex, age in years, and the success or failure of the treatment. The table below gives the probability for each combination of categories:

	Male		Female	
	Under 30	Over 30	Under 30	Over 30
Success	0·08	0·08	0·08	0·16
Failure	0·16	0·18	0·08	0·18

Show that there is the same degree of association between age of patient and result of treatment for both males and females, but that age and result are independent when the sexes are combined.

Construct a similar table where there is independence for both males and females, but a positive association between the categories of under 30 and success when the sexes are combined.

(Simpson, 1951; Birch, 1963)

56 The likelihood-ratio statistic for testing the hypothesis of zero two-response interaction is

$$Y^2 = 2(\Sigma_{ij}\, n_{ij} \log n_{ij} - \Sigma_i\, n_{io} \log n_{io} - \Sigma_j\, n_{oj} \log n_{oj} + n \log n).$$

A natural generalization for testing the hypothesis of zero three-response interaction is the statistic

$$L^2 = 2(\Sigma_{ijk}\, n_{ijk} \log n_{ijk} - \Sigma_{jk}\, n_{ojk} \log n_{ojk} - \Sigma_{ik}\, n_{iok} \log n_{iok}$$
$$- \Sigma_{ij}\, n_{ijo} \log n_{ijo} + \Sigma_i\, n_{ioo} \log n_{ioo} + \Sigma_j\, n_{ojo} \log n_{ojo} + \Sigma_k\, n_{ook} \log n_{ook}$$
$$- n \log n).$$

Show that L^2 converges to $\chi^2_{(r-1)\,(s-1)\,(t-1)}$ as $n \to \infty$ when H_ξ holds if and only if the three-way table is perfect, in the sense that

$$p_{ijk} = p_{ojk}\, p_{iok}\, p_{ijo}/p_{ioo}\, p_{ojo}\, p_{ook}.$$

57 A binary response R is observed at each treatment combination of two factors $\mathfrak{F}$ and $\mathfrak{G}$ which occur at 2 and t levels respectively. The category $i = 1$ represents success. At the kth level of $\mathfrak{G}$, the frequencies form the 2×2 table shown below.

Level of $\mathfrak{F}$	Success	Failure	Total
1	N_{11k}	N_{21k}	n_{01k}
2	N_{12k}	N_{22k}	n_{02k}
Total	N_{10k}	N_{20k}	n_{00k}

The mean and variance of N_{11k} conditional on the observed value of N_{10k} are given respectively by

$$m_k = n_{01k}n_{10k}/n_{00k} \quad \text{and} \quad v_k = n_{10k}n_{20k}n_{01k}n_{02k}/n_{00k}^2(n_{00k}-1)\ .$$

Write $m = \Sigma\, m_k$ and $v = \Sigma\, v_k$. Consider the hypothesis H_ξ, which specifies here that the interaction $\mathfrak{F} \times \mathfrak{G}$ has no effect on R. Show that an asymptotically valid test of H_ξ is obtained by referring the statistic

$$Q = \Sigma_k\,(N_{11k}-m_k)^2/v_k - (N_{110}-m)^2/v$$

to the distribution of χ^2_{t-1}.

(Zelen, 1971)

58 The experimental situation is the same as in the previous exercise. Suppose now that the interaction $\mathfrak{F} \times \mathfrak{G}$ has no effect on R. Write

$$\lambda = \log\,(p_{11k}p_{22k}/p_{12k}p_{21k}),$$

and define

$$U = (N_{110}-m)/v^{\frac{1}{2}}.$$

Show that the uniformly most powerful test of $\lambda \geqslant 0$ against $\lambda < 0$ is implemented for large n by referring U to the distribution $\mathfrak{N}(0, 1)$.

(Cochran, 1954; Mantel and Haenszel, 1959; Birch, 1964)

59 The conditions are the same as in Exercise 57, and H_ξ holds. Define

$$V = \Sigma_k\{(N_{11k}-m_k)/v_k^{\frac{1}{2}}\}.$$

Another test of $\lambda \geqslant 0$ against $\lambda < 0$ is known as the *summation of chi* procedure, and is implemented by referring V to the distribution $\mathfrak{N}(0, t)$. Criticize this procedure, and find its asymptotic relative efficiency as a test of H_λ when H_ξ holds.

(Yates, 1955b; Pasternack and Mantel, 1966; Snedecor and Cochran, 1967, p. 254; Armitage, 1971, p.373).

60 The conditions are the same as in the previous exercise. Suppose that a function $f(.)$ exists such that

$$f(p_{11k}) - f(p_{12k}) = \delta \quad \text{for all } k.$$

When $\delta = 0$, denote the common value of p_{11k} and p_{12k} by p_k. Introduce weights $w_1, w_2, \ldots, w_k$, and define

$$D = \Sigma_k w_k (N_{11k}/n_{01k} - N_{12k}/n_{02k}).$$

Consider a test of the hypothesis $\delta = 0$ against the alternative $\delta \neq 0$ based on the statistic $D/(\text{var } D)^{\frac{1}{2}}$. Show that the asymptotic efficiency of this test is maximized by taking

$$w_k \propto (dp/df)_{p=p_k} / \{p_k(1-p_k)n_{00k}/n_{01k}n_{02k}\}.$$

Deduce that the test of Exercise 58 is obtained by taking

$$f(p) = \log \{p/(1-p)\}.$$

(Radhakrishna, 1965)

61 A sample of size n is taken from the trivariate distribution of (R, S, T). The parameters $\{\xi_{abc}\}$ have specified values. Show that the maximum-likelihood estimate of np_{ijk} is $m_{ijk}(\boldsymbol{\xi})$. Prove also that

$$\Sigma_{ijk} \{n_{ijk} - m_{ijk}(\boldsymbol{\xi})\} \log m_{ijk}(\boldsymbol{\xi}) \equiv \Sigma_{abc}\{n_{abc} - m_{abc}(\boldsymbol{\xi})\}\xi_{abc}.$$

62 Consider the statistic K^2 defined in section 7.5. When $s = 2$ and $t = 3$, show that

$$K^2 = \sum_{k=1}^{3} A_k \{ \sum_{i=1}^{r} (n_{i20}\, n_{i1k} - n_{i10} n_{i2k})^2 / n_{i00} \} / (A_2 A_3 + A_3 A_1 + A_1 A_2),$$

where

$$A_1 = \sum_{i=1}^{r} n_{i10} n_{i20} n_{i02} n_{i03} / n_{i00}^2 (n_{i00} - 1),$$

and A_2, A_3 are defined by cyclic permutation.

(Birch, 1965).

63 An investigator classifies a random sample of diabetic patients in three ways: whether the disease started before the age of 45 ("early onset") or later; whether specified family members also had diabetes (+) or not (−); and whether the patient was dependent on insulin injections (+) or could be maintained with less inconvenient treatments (−). The results are shown in Table E.3.

Comment on the influence, if any, of family history on the other two responses and the relationship between them. If the circumstances are appropriate, obtain a point estimate and a confidence interval for the relative risk of being dependent on insulin for patients who have early onset as compared with late onset.

(D.J. Newell)

TABLE E.3 *Diabetic patients classified by time of onset, insulin dependence, and family history*

Family history	+		−	
Insulin dependence	+	−	+	−
Early onset	6	1	16	2
Late onset	6	36	8	48

64 The data in Table E.4 are quoted from Woolf (1955). Patients with peptic ulcer in London, Manchester, and Newcastle upon Tyne are classified

TABLE E.4 *Blood-groups of peptic ulcer and control patients in London, Manchester, and Newcastle upon Tyne*

Blood-group	London		Manchester		Newcastle	
	Control	Ulcer	Control	Ulcer	Control	Ulcer
A	4219	579	3775	246	5261	219
O	4578	911	4532	361	6598	396

by blood-group, and the same classification is made for control groups in the three cities. Analyse the dependence of type of patient on blood-group and location. Previous analyses are given by Woolf (*loc. cit.*), Ku and Kullback (1968), Berkson (1968), Cox (1970a, pp. 84-5), and Berkson (1972).

65 The data in Table E.5 are quoted from Kastenbaum and Lamphiear (1959). Litters of mice of different sizes are treated in one of two different ways, and the table gives the frequencies for 0, 1, and 2 or more depletions per litter. Analyse the dependence of depletions on litter size and treatment. Previous analyses are given by Kastenbaum and Lamphiear (*loc. cit.*), Plackett (1962), Goodman (1963b), Berkson (1968), Ku and Kullback (1968), Grizzle, Starmer and Koch (1969), and Berkson (1972).

66 A binary response is observed for each treatment combination of two factors $\mathfrak{F}$ and $\mathfrak{G}$. When the ith level of $\mathfrak{F}$ is combined with the jth level of $\mathfrak{G}$, the probability of success is given by

$$p_{ij1} = \mu + \alpha_i + \beta_j,$$

where

$$\Sigma\, \alpha_i = \Sigma\, \beta_j = 0.$$

Derive minimum modified chi-square estimates of μ, $\{\alpha_i\}$ and $\{\beta_j\}$. Hence obtain a test statistic for the hypothesis that the model satisfactorily fits the data. (Reiersøl, 1954)

TABLE E.5 *Litters of mice classified by litter size, treatment, and number of depletions*

Litter size	Treatment	Number of depletions 0	1	2 or more
7	*A*	58	11	5
	B	75	19	7
8	*A*	49	14	10
	B	58	17	8
9	*A*	33	18	15
	B	45	22	10
10	*A*	15	13	15
	B	39	22	18
11	*A*	4	12	17
	B	5	15	8

67 Table 7.1 presents data showing the effect of time of planting and length of cutting on the propagation of plum root-stocks. The analysis in section 7.6 assumes that the differences between probabilities are measured on a logit scale. Snedecor and Cochran (1967, pp. 495-6) give an analysis based on a linear scale. Calculate the maximum likelihood estimates of the expected frequencies for the model defined in the previous exercise. The estimated frequencies for live shoots given below are quoted from Berkson (1972).

At once		In spring	
Long	Short	Long	Short
82·889	134·212	157·113	208·436

68 Two treatments are given to one group in the order $(\mathfrak{F}, \mathfrak{G})$ and to another group in the order $(\mathfrak{G}, \mathfrak{F})$. A binary response is observed. In the first group, the frequency of untied pairs is G_1, and the frequency where $\mathfrak{F}$ leads to success and $\mathfrak{G}$ to failure is F_1. The corresponding frequencies in the second group are G_2 and F_2. Conditional on the observed value g_i, F_i has a binomial distribution with index g_i and probability $(1 + e^{-\nu_i})^{-1}$ for $i = 1, 2$. Write

$$\nu_1 = \epsilon + \delta \quad \text{and} \quad \nu_2 = -\epsilon + \delta,$$

where ϵ is the order effect and δ is the treatment effect. Show that the statistics $F_1 + F_2$ and $F_1 - F_2$ are jointly sufficient for ϵ and δ. Deduce that tests of $\delta = 0$ and $\epsilon = 0$ can be based on the 2×2 tables shown below

Test of $\delta = 0$ eliminating ϵ		Test of $\epsilon = 0$ eliminating δ	
F_1	$g_2 - F_2$	F_1	F_2
$g_1 - F_1$	F_2	$g_1 - F_1$	$g_2 - F_2$

(Gart, 1969)

69 In a clinical trial to compare two analgesics, A and B, 175 patients are randomly allocated to one of the four sequences AB, BA, AA, BB. The pairs of letters indicate the treatments the patient receives in the first and second periods of the trial, respectively. Each patient was asked to express a preference for the first or second treatment that he received, with the following results:

Sequence	Prefers first	Prefers second	No preference
AB	16	4	20
BA	8	12	22
AA	10	5	30
BB	11	6	31

Analyse the data from the AB and BA sequences to test whether there is evidence of a treatment effect in the presence of a possible period effect, and vice versa.

Analyse the remainder of the data for period effect, and propose a method of combining the two analyses for period effect.

(D.J. Newell)

70 A matched pairs design is used to compare two treatments. The observations are binary variates T_1 and T_2. Consider a model in which

(i) for every pair, $p(T_1 = u, T_2 = v) = \theta_{uv}$ $(u,v = 0, 1)$;
(ii) the parameters $\{\theta_{uv}\}$ are the observed values of variates $\{\Theta_{uv}\}$;
(iii) the prior distribution of $\{\Theta_{uv}\}$ has the Dirichlet form with parameters $\{\beta'_{uv}\}$.

Let f_{uv} be the observed frequency of the event $(T_1 = u, T_2 = v)$ and define

$$\beta_{uv} = \beta'_{uv} + f_{uv}\,.$$

Show that

$$p(\Theta_{10} > \Theta_{01}) = p(W > \tfrac{1}{2}),$$

where the probability on the left is calculated from the posterior distribution of $\{\Theta_{uv}\}$, and the variate W has a beta distribution with density function proportional to $w^{\beta_{01}-1}(1-w)^{\beta_{10}-1}$.

(Altham, 1971a)

71 Table 9.3 shows, for a three-way classification, the relationship between the parameters which are zero in a hierarchical model and the marginal totals which are fixed. Construct a similar table for a four-way classification.

(Goodman, 1970)

72 The marginal totals which are fixed in a multidimensional contingency table consist of $\{N_{u_1}\}$, $\{N_{u_2}\}$, . . . , $\{N_{u_c}\}$, where $u_1, u_2, \ldots, u_c$ consist of subsets of $i, j, k, \ldots, l$ together with zeros. When $u_1, u_2, \ldots, u_c$ have no letters in common, show that the maximum-likelihood estimators of the expected frequencies for the corresponding hierarchical model are given by

$$m_{ijk\ldots l} = N_{u_1} N_{u_2} \ldots N_{u_c} / bN^{c-1},$$

where b is the product of the number of categories for each response with a letter not contained in any of $u_1, u_2, \ldots, u_c$.

When there is only one set of letters, say w, that is contained in two or more of $u_1, u_2, \ldots, u_c$, show that

$$m_{ijk\ldots l} = N_{u_1} N_{u_2} \ldots N_{u_c} / bN^{a} N_{w}^{a'-1},$$

where a is the number of subsets $u_1, u_2, \ldots, u_c$ that do not contain w, and a' is the number that do.

Hence obtain the expected frequencies for all the models in Table 9.3 except the first.

(Goodman, 1970)

73 A new detergent X and a standard detergent M were each placed with 1008 consumers selected at random. After a period, consumers were classified with respect to the following:

(i) the product preferred (X/M),

(ii) the detergent generally used (M/not-M),

(iii) the kind of water used (hard/medium/soft),

(iv) the temperature of water used (101° to 115°/over 115°F).

The data are given in Table E.6, which is quoted from Ries and Smith (1963).

TABLE E.6 *Classification of 1008 consumers by water softness, previous use of brand M, water temperature, and preference for brand X*

Water softness	Brand preference	Previous user of M High temp.	Previous user of M Low temp.	Previous non-user of M High temp.	Previous non-user of M Low temp.
Soft	X	19	57	29	63
	M	29	49	27	53
Medium	X	23	47	33	66
	M	47	55	23	50
Hard	X	24	37	42	68
	M	43	52	30	42

Their objective was to find the relationship between the product preferred and the other factors in the study. Analyse the data, (a) assuming that preference for detergent brand X over brand M is a response and the other variables are factors, (b) assuming that each variable is a response. Previous analyses are given by Ries and Smith (*loc. cit.*), Cox and Lauh (1967), Ku and Kullback (1968), Cox (1970a, pp. 37-40), Goodman (1971), and Johnson and Koch (1971).

74 Individuals were classified according to the presence or absence of newspapers, radio, solid reading, lectures, and a good score in a test of cancer knowledge. The data in Table E.7 are quoted from Lombard and Doering (1947).

TABLE E.7 *Classification of 1729 individuals according to presence or absence of newspapers, radio, solid reading, lectures, and a good score in a test of cancer knowledge*

	Radio				No radio			
	Solid reading		No solid reading		Solid reading		No solid reading	
	Newspapers							
Knowledge	Good	Poor	Good	Poor	Good	Poor	Good	Poor
Lectures	23	8	8	4	27	18	7	6
No lectures	102	67	35	59	201	177	75	156
	No Newspapers							
Knowledge	Good	Poor	Good	Poor	Good	Poor	Good	Poor
Lectures	1	3	4	3	3	8	2	10
No lectures	16	16	13	50	67	83	84	393

Analyse the data (a) assuming that cancer knowledge is a response and the other variables are factors, (b) assuming that each variable is a response. Previous analyses are given by Lombard and Doering (*loc. cit.*), Winsor (1948), Dyke and Patterson (1952), Yates (1960, sect. 9.7), Cox and Snell (1968), Bishop (1969), and Goodman (1970).

75 New Jersey high-school seniors were given an attitude questionnaire in 1957 entitled "Attitudes toward science and scientific careers". In addition, each student obtained a score on a brief IQ vocabulary test. The score was used to divide the 2982 individuals into two groups of 1491, namely high IQ and low IQ. Only four of the many items on the questionnaire are used here. The data are given in Table E.8.

TABLE E.8 *Classification of 2982 individuals according to IQ and four binary responses*

Response	Low IQ	High IQ	Response	Low IQ	High IQ
1111	62	122	0111	14	20
1110	70	68	0110	11	10
1101	31	33	0101	11	11
1100	41	25	0100	14	9
1011	283	329	0011	31	56
1010	253	247	0010	46	55
1001	200	172	0001	37	64
1000	305	217	0000	82	53

Reprinted with permission of the publisher from "Classification Procedures based on Dichotomous Response Vectors", by Herbert Solomon, in *Studies in Item Analysis and Prediction*, edited by Herbert Solomon (Stanford: Stanford University Press, 1961), page 180.

An observation 1 means agreement and 0 means disagreement with the following assertions, which correspond respectively to the four digits in the response column of the table:

(i) The development of new ideas is the scientist's greatest source of satisfaction.

(ii) Scientists and engineers should be eliminated from the military draft.

(iii) The scientist will make his maximum contribution to society when he has freedom to work on problems which interest him.

(iv) The monetary compensation of a Nobel Prize winner in physics should be at least equal to that given popular entertainers.

Analyse the relationship between the four responses in each IQ group, and compare the two groups. Previous analyses are given by Solomon (*loc. cit.*), Bock (1970), and Martin and Bradley (1972).

REFERENCES

Aitchison, J. & Silvey, S.D., 1957, The generalization of probit analysis to the case of multiple responses. *Biometrika*, **44**, 131–40.

Ajne, B., 1968, A simple test for uniformity of a circular distribution. *Biometrika*, **55**, 343–54.

Altham, Patricia M.E., 1969, Exact Bayesian analysis of a 2 × 2 contingency table and Fisher's "exact" significance test. *J.R. Statist. Soc. B*, **31**, 261–9.

Altham, Patricia M.E., 1970a, The measure of association of rows and columns in a $r \times s$ contingency table. *J.R. Statist. Soc. B*, **32**, 63–73.

Altham, Patricia M.E., 1970b, The measurement of association in a contingency table: three extensions of the cross-ratios and metrics methods. *J.R. Statist. Soc. B*, **32**, 395–407.

Altham, Patricia M.E., 1971a, The analysis of matched proportions. *Biometrika*, **58**, 561–76.

Altham, Patricia M.E., 1971b, Exact Bayesian analysis of an intraclass 2 × 2 table. *Biometrika*, **58**, 679–80.

Anderson, J.A., 1972, Separate sample logistic discrimination. *Biometrika*, **59**, 19–35.

Anderson, R.L., 1959, Use of contingency tables in the analysis of consumer preference studies. *Biometrics*, **15**, 582–90.

Anderson, T.W., 1958, *An Introduction to Multivariate Statistical Analysis*. New York: Wiley.

Anscombe, F.J., 1948, The transformation of Poisson, binomial and negative-binomial data. *Biometrika*, **35**, 246–54.

Anscombe, F.J., 1956, On estimating binomial response relations. *Biometrika*, **43**, 461–4.

Armitage, P., 1955, Tests for linear trends in proportions and frequencies. *Biometrics*, **11**, 375–86.

Armitage, P., 1971, *Statistical Methods in Medical Research*. Oxford & Edinburgh: Blackwell.

Armitage, P. & Healy, M.J.R., 1957, Interpretation of χ^2 tests. *Biometrics*, **13**, 113–15.

Ashford, J.R., 1959, An approach to the analysis of data for semi-quantal responses in biological assay. *Biometrics*, **15**, 573–81.

Ashford, J.R. & Sowden, R.R., 1970, Multivariate probit analysis. *Biometrics*, **26**, 535–46.

Ashton, Winifred D., 1972, *The Logit Transformation with special reference to its uses in bioassay*. London : Griffin.

Assakul, K. & Proctor, C.H., 1967, Testing independence in two-way contingency tables with data subject to misclassification. *Psychometrika*, **32**, 67–76.

Barlow, R.E., Bartholomew, D.J., Bremner, J.M. & Brunk, H.D., 1972, *Statistical Inference under Order Restrictions: the theory and applications of isotonic regression*. New York : Wiley.

Barnard, G.A., 1947, Significance tests for 2 × 2 tables. *Biometrika*, **34**, 123–38.

Bartlett, M.S., 1935, Contingency table interactions. *Suppl. J.R. Statist. Soc.*, **2**, 248–52.

Bartlett, M.S., 1937, Properties of sufficiency and statistical tests. *Proc. Roy. Soc. A*, **160**, 268–82.

Barton, D.E., 1961, Unbiased estimation of a set of probabilities. *Biometrika*, **48**, 227–9.

Bateman, G.I., 1950, The power of the χ^2 index of dispersion test when Neyman's contagious distribution is the alternate hypothesis. *Biometrika*, **37**, 59–63.

Bennett, B.M., 1959, A sampling study of the power function of the χ^2 "index of dispersion" test. *J. Hyg., Camb.*, **57**, 360–6.

Bennett, B.M., 1967, Tests of hypotheses concerning matched samples. *J.R. Statist. Soc. B*, **29**, 468–74; corr., **31**, 194.

Bennett, B.M., 1968, Note on X^2 tests for matched samples. *J.R. Statist. Soc. B*, **30**, 368–70.

Berkson, J., 1944, Application of the logistic function to bio-assay. *J. Amer. Statist. Assoc.*, **39**, 357–65.

Berkson, J., 1955, Maximum likelihood and minimum χ^2 estimation of the logistic function. *J. Amer. Statist. Assoc.*, **50**, 130–62.

Berkson, J., 1968, Application of minimum logit χ^2 estimate to a problem of Grizzle with a notation on the problem of "no interaction". *Biometrics*. **24**, 75–96.

Berkson, J., 1972, Minimum discrimination information, the "no interaction" problem, and the logistic function. *Biometrics*, **28**, 443–68.

Bhapkar, V.P., 1966, A note on the equivalence of two test criteria for hypotheses in categorical data. *J. Amer. Statist. Assoc.*, **61**, 228–35.

Bhapkar, V.P. & Koch, G.G., 1968, Hypothesis of "no interaction" in multidimensional contingency tables. *Technometrics*, **10**, 107–23.

Billewicz, W.Z., 1965, The efficiency of matched samples: an empirical investigation. *Biometrics*, **21**, 623–44.

Birch, M.W., 1963, Maximum likelihood in three-way contingency tables. *J.R. Statist. Soc. B*, **25**, 220–33.

Birch, M.W., 1964, The detection of partial association, I: the 2 × 2 case. *J.R. Statist. Soc. B*, **26**, 313–24.

Birch, M.W., 1965, The detection of partial association, II: the general case. *J.R. Statist. Soc. B*, **27**, 111–24.

Bishop, Yvonne M.M., 1969, Full contingency tables, logits, and split contingency tables. *Biometrics*, **25**, 383–99.

Bishop, Yvonne M.M., 1971, Effects of collapsing multidimensional contingency tables. *Biometrics*, **27**, 545–62.

Bishop, Yvonne M.M. & Fienberg, S.E., 1969, Incomplete two-dimensional contingency tables. *Biometrics*, **25**, 119–28.

Bloch, D.A. & Watson, G.S., 1967, A Bayesian study of the multinomial distribution. *Ann. Math. Statist.*, **38**, 1423–35.

Bloomfield, P., 1972, *Transformations for multivariate binary data.* Tech. Rep. 18, Ser. 2, Dept. Statist., Princeton Univ.

Bock, R.D., 1970, Estimating multinomial response relations, pp. 111–32 in R.C. Bose *et al.* (editors), *Essays in Probability and Statistics.* Chapel Hill : University of North Carolina Press.

Bowker, A.H., 1948, A test for symmetry in contingency tables. *J. Amer. Statist. Assoc.*, **43**, 572–4.

Bross, I., 1954, Misclassification in 2 × 2 tables. *Biometrics*, **10**, 478–86.

Carothers, A.D., 1971, An examination and extension of Leslie's test of equal catchability. *Biometrics*, **27**, 615–30.

Caussinus, H., 1965, Contribution à l'analyse statistique des tableaux de correlation. *Ann. Fac. Sci. Univ. Toulouse*, **29**, 77–182.

Chakravarti, I.M. & Rao, C.R., 1959, Tables for some small sample tests of significance for Poisson distributions and 2 X 3 contingency tables. *Sankhyā*, **21**, 315–26.

Chambers, J.M., 1973, Fitting nonlinear models : numerical techniques. *Biometrika*, **60**, 1–13.

Charles, Enid, 1953, The hour of birth. *Brit. J. Prev. Soc. Med.*, **7**, 43–59.

Chase, G.R., 1968, On the efficiency of matched pairs in Bernoulli trials. *Biometrika*, **55**, 365–9.

Cochran, W.G., 1936, The χ^2 distribution for the binomial and Poisson series, with small expectations. *Ann. Eugen. Lond.*, **7**, 207–17.

Cochran, W.G., 1950, The comparison of percentages in matched samples. *Biometrika*, **37**, 256–66.

Cochran, W.G., 1952, The χ^2 test of goodness of fit. *Ann. Math. Statist.*, **23**, 315–45.

Cochran, W.G., 1954, Some methods for strengthening the common χ^2 tests. *Biometrics*, **10**, 417–51.

Cochran, W.G. & Hopkins, C.E., 1961, Some classification problems with multivariate qualitative data. *Biometrics*, **17**, 10–32.

Cornfield, J., 1956, A statistical problem arising from retrospective studies. *Proc. 3rd Berkeley Symp. Math. Statist. and Prob.* (ed. J. Neyman), Vol. 4, pp.135–48.

Cox, D.R., 1958, The regression analysis of binary sequences (with Discussion). *J.R. Statist. Soc. B*, **20**, 215–42.

Cox, D.R., 1970a, *The Analysis of Binary Data*. London : Methuen.

Cox, D.R., 1970b, The continuity correction. *Biometrika*, **57**, 217–19.

Cox, D.R., 1972, The analysis of multivariate binary data. *Appl. Statist.*, **21**, 113–20.

Cox, D.R. & Lauh, E., 1967, A note on the graphical analysis of multidimensional contingency tables. *Technometrics*, **9**, 481–8.

Cox, D.R. & Snell, E.J., 1968, A general definition of residuals (with Discussion). *J.R. Statist. Soc. B*, **30**, 248–75.

Craddock, J.M. & Flood, C.R., 1970, The distribution of the χ^2 statistic in small contingency tables. *Appl. Statist.*, **19**, 173–81.

Craig, C.C., 1953, Combination of neighboring cells in contingency tables. *J. Amer. Statist. Assoc.*, **48**, 104–12.

Cramér, H., 1946, *Mathematical Methods of Statistics*. Princeton University Press.

Daniel, C., 1959, Use of half-normal plots in interpreting factorial two-level experiments. *Technometrics*, **1**, 311–41.

Darroch, J.N., 1962, Interactions in multi-factor contingency tables. *J.R. Statist. Soc. B*, **24**, 251–63.

Darroch, J.N., 1971, A definition of independence for bounded-sum, nonnegative, integer-valued variables. *Biometrika*, **58**, 357–68.

Darroch, J.N. & Ratcliff, D., 1972, Generalized iterative scaling for log-linear models. *Ann. Math. Statist.*, **43**, 1470–80.

Darroch, J.N. & Ratcliff, D., 1973, Tests of *F*-independence with reference to quasi-independence and Waite's fingerprint data. *Biometrika*, **60**, 395–401.

Darwin, J.H., 1957, The power of the Poisson index of dispersion. *Biometrika*, **44**, 286–9.

David, H.A. & Newell, D.J., 1965, The identification of annual peak periods for a disease. *Biometrics*, **21**, 645–50.

Dawson, R.B., 1954, A simplified expression for the variance of the χ^2-function on a contingency table. *Biometrika*, **41**, 280.

Draper, N.R. & Smith, H., 1966, *Applied Regression Analysis*. New York : Wiley.

Drinkwater, R.W. & Hastings, N.A.J., 1967, An economic replacement model. *Opl. Res. Q.*, **18**, 121–38.

Dyke, A.V. & Patterson, H.D., 1952, Analysis of factorial arrangements when the data are proportions. *Biometrics*, **8**, 1–12.

Edwards, A.W.F., 1963, The measure of association in a 2 × 2 table. *J.R. Statist. Soc. A*, **126**, 109–14.

Edwards, A.W.F. & Fraccaro, M., 1960, Distribution and sequences of sexes in a selected sample of Swedish families. *Ann. Human Genet., Lond.*, **24**, 245–52.

Elderton, Ethel M., 1920, On the inheritance of the finger-print. *Biometrika*, **13**, 57–91.

Evans, D.A., 1973, The influence of computers on the teaching of statistics. *J.R. Statist. Soc. A*, **136**, 153–90.

Feller, W., 1968, *An Introduction to Probability Theory and its Applications*, Vol. 1, 3rd edn. New York : Wiley.

Fienberg, S.E., 1970a, Quasi-independence and maximum likelihood estimation in incomplete contingency tables. *J. Amer. Statist. Assoc.*, **65**, 1610–16.

Fienberg, S.E., 1970b, An iterative procedure for estimation in contingency tables. *Ann. Math. Statist.*, **41**, 907–17.

Fienberg, S.E., 1972, The analysis of incomplete multi-way contingency tables. *Biometrics*, **28**, 177–202 (corr., **29**, 829).

Fienberg, S.E. & Holland, P.W., 1970, Methods for eliminating zero counts in contingency tables, pp. 233–60 in G.P. Patil (ed.), *Random Counts in Scientific Work* : Vol. 1, *Random Counts in Models and Structures*. Pennsylvania State University Press.

Finney, D.J., 1971, *Probit Analysis*, 3rd edn. Cambridge: University Press.

Finney, D.J., Latscha, R., Bennett, B.M. & Hsu, P., with an introduction by Pearson, E.S., 1963, *Tables for Testing Significance in a 2 × 2 Contingency Table*. Cambridge : University Press.

Fisher, R.A., 1934, *Statistical Methods for Research Workers*, 5th edn. Edinburgh : Oliver & Boyd (14th edn, 1970).

Fisher, R.A., 1935, The logic of inductive inference (with Discussion). *J.R. Statist. Soc.*, **98**, 39–54.

Fisher, R.A., 1942, The likelihood solution of a problem in compounded probabilities. *Ann. Eugen. Lond.*, **11**, 306–7.

Fisher, R.A., 1950, The significance of deviations from expectation in a Poisson series. *Biometrics*, **6**, 17–23; **20**, 265–72.

Fisher, R.A., 1959, *Statistical Methods and Scientific Inference*, 2nd edn. Edinburgh : Oliver & Boyd.

Fisher, R.A., 1962, Confidence limits for a cross-product ratio. *Austral. J. Statist.*, **4**, 41.

Fisher, R.A., Thornton, H.G. & Mackenzie, W.A., 1922, The accuracy of the plating method of estimating the density of bacterial populations. *Ann. Appl. Biol.*, **9**, 325–59.

Fix, E., Hodges, J.L. & Lehmann, E.L., 1959, The restricted chi-square test, pp. 92–107 in U. Grenander (ed.), *Probability and Statistics : the Harald Cramér Volume*. Stockholm : Almqvist & Wiksell.

Freeman, G.H. & Halton, J.H., 1951, Note on an exact treatment of contingency, goodness of fit and other problems of significance. *Biometrika*, **38**, 141–9.

Freeman, M.F. & Tukey, J.W., 1950, Transformations related to the angular and square root. *Ann. Math. Statist.*, **21**, 607–11.

Gabriel, K.R., 1966, Simultaneous test procedures for multiple comparisons on categorical data. *J. Amer. Statist. Assoc.*, **61**, 1081–96.

Gart, J.J., 1962, Approximate confidence limits for the relative risk. *J.R. Statist. Soc. B*, **24**, 454–63.

Gart, J.J., 1966, Alternative analyses of contingency tables. *J.R. Statist. Soc. B*, **28**, 164–79.

Gart, J.J., 1969, An exact test for comparing matched proportions in cross-over designs. *Biometrika*, **56**, 75–80.

Gart, J.J., 1970, Point and interval estimation of the common odds ratio in the combination of 2 × 2 tables with fixed marginals. *Biometrika*, **57**, 471–5.

Gart, J.J., 1972, Interaction tests for 2 × s × t contingency tables. *Biometrika*, **59**, 309–16.

Gart, J.J. & Thomas, D.G., 1972, Numerical results on approximate confidence limits for the odds ratio. *J.R. Statist. Soc. B.*, **34**, 441–7.

Gart, J.J. & Zweifel, J.R., 1967, On the bias of various estimators of the logit and its variance with application to quantal bioassay. *Biometrika*, **54**, 181–7.

Glasser, C.J., 1962, Minimum variance unbiased estimators for Poisson probabilities. *Technometrics*, **4**, 409–18.

Gokhale, D.V., 1972, Analysis of log-linear models. *J.R. Statist. Soc. B*, **34**, 371–6.

Good, I.J., 1967, A Bayesian significance test for multinomial distributions (with Discussion). *J.R. Statist. Soc. B*, **29**, 399–431.

Good, I.J., Gover, T.N. & Mitchell, G.J., 1970, Exact distributions for χ^2 and for the likelihood-ratio statistic for the equiprobable multinomial distribution. *J. Amer. Statist. Assoc.*, **65**, 267–83.

Goodman, L.A., 1963a, Statistical methods for the preliminary analysis of transaction flows. *Econometrica*, **31**, 197–208.

Goodman, L.A., 1963b, On Plackett's test for contingency table interactions. *J.R. Statist. Soc. B*, **25**, 179–88.

Goodman, L.A., 1964, Simultaneous confidence limits for cross-product ratios in contingency tables. *J.R. Statist. Soc. B*, **26**, 86–102.

Goodman, L.A., 1965, On simultaneous confidence intervals for multinomial proportions. *Technometrics*, **7**, 247–54.

Goodman, L.A., 1968, The analysis of cross-classified data: independence, quasi-independence, and interactions in contingency tables with or without missing entries. *J. Amer. Statist. Assoc.*, **63**, 1091–1131.

Goodman, L.A., 1969, On partitioning χ^2 and detecting partial association in three-way contingency tables. *J.R. Statist. Soc. B*, **31**, 486–98.

Goodman, L.A., 1970, The multivariate analysis of qualitative data: interactions among multiple classifications. *J. Amer. Statist. Assoc.*, **65**, 226–56.

Goodman, L.A., 1971, The analysis of multidimensional contingency tables: stepwise procedures and direct estimation methods for building models for multiple classifications. *Technometrics*, **13**, 33–61.

Goodman, L.A. & Kruskal, W.H., 1954, 1959, 1963, 1972, Measures of association for cross classifications, Parts I, II, III, IV. *J. Amer. Statist. Assoc.*, **49**, 732–64 (corr. 52, 578); 54, 123–63; 58, 310–64; 67, 415–21.

Greenwood, M. & Yule, G.U., 1920, An inquiry into the nature of frequency distributions representative of multiple happenings with particular reference to the occurrence of multiple attacks of disease or of repeated accidents. *J.R. Statist. Soc.*, **83**, 255–79.

Grizzle, J.E., Starmer, C.F. & Koch, G.G., 1969, Analysis of categorical data by linear models. *Biometrics*, **25**, 489–504.

Haberman, S.J., 1970, *The general log-linear model*. Univ. Chicago, Ph.D. dissertation.

Haberman, S.J., 1972, Log-linear fit for contingency tables. *Appl. Statist.*, **21**, 218–25.

Haberman, S.J., 1973a, Printing multidimensional tables. *Appl. Statist.*, **22**, 118–26.

Haberman, S.J., 1973b, The analysis of residuals in cross-classified tables. *Biometrics*, **29**, 205–20.

Haberman, S.J., 1973c, *The Analysis of Frequency Data*. IMS Monograph Series.

Haight, F., 1967, *Handbook of the Poisson Distribution*. New York : Wiley.

Haldane, J.B.S., 1937, The exact value of the moments of the distribution of χ^2, used as a test of goodness of fit when expectations are small. *Biometrika*, **29**, 133–43. (corr., **31**, 220).

Haldane, J.B.S., 1940, The mean and variance of χ^2, when used as a test of homogeneity, when expectations are small. *Biometrika*, **31**, 346–55.

Haldane, J.B.S., 1955, The estimation and significance of the logarithm of a ratio of frequencies. *Ann. Human Genet., Lond.*, **20**, 309–11.

Hannan, J. & Harkness, W.L., 1963, Normal approximation to the distribution of two independent binomials conditional on a fixed sum. *Ann. Math. Statist.*, **34**, 1593–5.

Harkness, W.L., 1965, Properties of the extended hypergeometric distribution. *Ann. Math. Statist.*, **36**, 938–45.

Haynam, G.E. & Leone, F.C., 1965, Analysis of categorical data. *Biometrika*, **52**, 654–60.

Healy, M.J.R., 1969, Exact tests of significance in contingency tables. *Technometrics*, **11**, 393–5.

Hewlett, P.S. & Plackett, R.L., 1950, Statistical aspects of the independent joint action of poisons, particularly insecticides: II. Examination of data for agreement with the hypothesis. *Ann. Appl. Biol.*, **37**, 527–52.

Hills, M., 1966, Allocation rules and their error rates (with Discussion). *J.R. Statist. Soc. B*, **28**, 1–31.

Hills, M., 1967, Discrimination and allocation with discrete data. *Appl. Statist.*, **16**, 237–50.

Hinz, P. & Gurland, J., 1968, A method of analysing untransformed data from the negative binomial and other contagious distributions. *Biometrika*, **55**, 163–70.

Hoblyn, T.N. & Palmer, R.C., 1934, A complex experiment in the propagation of plum rootstocks from root cuttings. *J. Pomology & Hort. Sci.*, **12**, 36–56.

Hoel, P.G., 1943, On indices of dispersion. *Ann. Math. Statist.*, **14**, 155–62.

Holmes, M.C. & Williams, R.E.O., 1954, The distribution of carriers of *Streptococcus pyogenes* among 2,413 healthy children. *J. Hyg., Camb.*, **52**, 165–79.

Hoyt, C.J., Krishnaiah, P.R. & Torrance, E.P., 1959, Analysis of complex contingency data. *J. Exp. Ed.*, **27**, 187–94.

Innes, J.R.M., Ulland, B.M., Valerio, Marian G., Petrucelli, L., Fishbein, L., Hart, E.R., Pallotta, A.J., Bates, R.R., Falk, H.L., Gart, J.J., Klein, M., Mitchell, I. & Peters, J., 1969, Bioassay of pesticides and industrial chemicals for tumorigenicity in mice : a preliminary note. *J. Nat. Cancer Inst.*, **42**, 1101–14.

Ireland, C.T., Ku, H.H. & Kullback, S., 1969, Symmetry and marginal homogeneity of an $r \times r$ contingency table. *J. Amer. Statist. Assoc.*, **64**, 1323–41.

Ireland, C.T. & Kullback, S., 1968, Contingency tables with given marginals. *Biometrika*, **55**, 179–88.

Irwin, J.O., 1935, Tests of significance for differences between percentages based on small numbers. *Metron*, **12**, 83–94.

Johnson, W.D. & Koch, G.G., 1971, A note on the weighted least squares analysis of the Ries–Smith contingency table data. *Technometrics*, **13**, 438–47.

Kalbfleisch, J.G. & Sprott, D.A., 1973, The comparison of Poisson-distributed observations. *Biometrics*, **29**, 223–4.

Kastenbaum, M.A. & Lamphiear, D.E., 1959, Calculation of chi-square to test the no three-factor interaction hypothesis. *Biometrics*, **15**, 107–15.

Kendall, M.G. & Stuart, A., 1958 on. *The Advanced Theory of Statistics*, 3 vol. London: Griffin.

Kihlberg, J.K., Narragon, E.A. & Campbell, B.J., 1964, *Automobile crash injury in relation to car size*. Cornell Aero. Lab. Rep. No. VJ–1823–R11.

Ku, H.H. & Kullback, S., 1968, Interaction in multidimensional contingency tables: an information theoretic approach. *J. Res. Nat. Bur. Stds, B*, **72**, 159–99.

Lancaster, H.O., 1949, The derivation and partition of χ^2 in certain discrete distributions. *Biometrika*, **36**, 117–29.

Lancaster, H.O., 1967, Discussion of F. Mosteller (1967), Statistical comparisons of anesthetics : the national halothane study. *Bull. Int. Statist. Inst.*, **42**, 438–40.

Lancaster, H.O., 1969a, *The Chi-squared Distribution*, New York : Wiley.

Lancaster, H.O., 1969b, Contingency tables of higher dimensions. *Bull. Int. Statist. Inst.*, **43**, 143–51.

Lancaster, H.O. & Brown, I.A.I., 1965, Sizes of the χ^2 test in the symmetric multinomials. *Austral. J. Statist.*, 7, 40–44.

Lee, J.A.H., 1963, Seasonal variation in leukaemia incidence. *Brit. Med. J.*, (II), 623.

Lehmann, E.L., 1959, *Testing Statistical Hypotheses.* New York : Wiley.

Lewontin, R.C. & Felsenstein, J., 1965, The robustness of homogeneity tests in 2 × *N* tables. *Biometrics*, **21**, 19–33.

Lindley, D.V., 1964, The Bayesian analysis of contingency tables. *Ann. Math. Statist.*, **35**, 1622–43.

Lindley, D.V., 1965, *Introduction to Probability and Statistics from a Bayesian Viewpoint*. Part 2: Inference. Cambridge: University Press.

Lombard, H.L. & Doering, C.R., 1947, Treatment of the fourfold table by partial association and partial correlation as it relates to public health problems. *Biometrics*, **3**, 123–8.

McNemar, Q., 1947, Note on the sampling error of the difference between correlated proportions or percentages. *Psychometrika*, **12**, 153–7.

Madansky, A., 1963, Tests of homogeneity for correlated samples. *J. Amer. Statist. Assoc.*, **58**, 97–119.

Mantel, N., 1963, Chi-square tests with one degree of freedom; extensions of the Mantel–Haenszel procedure. *J. Amer. Statist. Assoc.*, **58**, 690–700.

Mantel, N., 1966, Models for complex contingency tables and polychotomous dosage response curves. *Biometrics*, **22**, 83–95.

Mantel, N.& Brown, C., 1973, A logistic reanalysis of Ashford and Sowden's data on respiratory symptoms in British coal miners. *Biometrics*, **29**, 649–65.

Mantel, N. & Haenszel, W., 1959, Statistical aspects of the analysis of data from retrospective studies of disease. *J. Nat. Cancer Inst.*, **22**, 719–48.

Mantel, N. & Hankey, B.F., 1971, Programmed analysis of a 2 × 2 contingency table. *Amer. Statistician*, **25** (5), 40–44.

Mardia, K.V., 1972, *Statistics of Directional Data*. London : Academic Press.

Martin, D.C. & Bradley, R.A., 1972, Probability models, estimation, and classification for multivariate dichotomous populations. *Biometrics*, **28**, 203–21.

Mather, K., 1940, The design and significance of synergic action tests. *J. Hyg., Camb.*, **40**, 513–31.

Matui, I., 1932, Statistical study of the distribution of scattered villages in two regions of the Tonami Plain, Toyama prefecture. *Jap. J. Geol. Geog.*, **9**, 251–66.

Maxwell, A.E., 1961, *Analysing Qualitative Data*. London : Methuen.

Meng, Rosa C. & Chapman, D.G., 1966, The power of chi-square tests for contingency tables. *J. Amer. Statist. Assoc.*, **61**, 965–75.

Miettinen, O.S., 1968, The matched pairs design in the case of all-or-none responses. *Biometrics*, **24**, 339–52.

Mitra, S.K., 1958, On the limiting power function of the frequency chi-square test. *Ann. Math. Statist.*, **29**, 1221–33.

Mosteller, F., 1952, Some statistical problems in measuring the subjective response to drugs. *Biometrics*, **8**, 220–6.

Mote, V.L. & Anderson, R.L., 1965, An investigation of the effect of misclassification on the properties of χ^2-tests in the analysis of categorical data. *Biometrika*, **52**, 95–109.

Nam, J.-M., 1971, On two tests for comparing matched proportions. *Biometrices*, **27**, 945–59.

Nelder, J.A. & Wedderburn, R.W.M., 1972, Generalized linear models. *J.R. Statist. Soc. A*, **135**, 370–84.

Newell, D.J., 1963, Misclassification in 2 × 2 tables. *Biometrics*, **19**, 187–8.

Neyman, J., 1949, Contribution to the theory of the χ^2-test. *Proc. 1st Berkeley Symposium on Math. Statist. and Prob.*, pp. 239–73. Berkeley : University of California Press.

Nicholson, W.L., 1956, On the normal approximation to the hypergeometric distribution. *Ann. Math. Statist.*, **27**, 471–83.

Odoroff, C.L., 1970, A comparison of minimum logit chi-square estimation and maximum likelihood estimation in 2 × 2 × 2 and 3 × 2 × 2 contingency tables: tests for interaction. *J. Amer. Statist. Assoc.*, **65**, 1617–31.

Okamoto, M., 1963, Chi-square statistic based on the pooled frequencies of several observations. *Biometrika*, **50**, 524–8.

Okamoto, M. & Ishii, G., 1961, Test of independence in intraclass 2 × 2 tables. *Biometrika*, **48**, 181–90.

Ord, J.K., 1972, *Families of Frequency Distributions*. London : Griffin.

Pasternack, B.S. & Mantel, N., 1966, A deficiency in the summation of chi procedure. *Biometrics*, **22**, 407–9.

Patil, G.P., 1963, Minimum variance unbiased estimation and certain problems of additive number theory. *Ann. Math. Statist.*, **34**, 1050–6.

Patnaik, P.B., 1949, The non-central χ^2 and F-distributions and their applications. *Biometrika*, **36**, 202–32.

Pearson, E.S., 1947, The choice of statistical tests illustrated on the interpretation of data classed in a 2 × 2 table. *Biometrika*, **34**, 139–67.

Pearson, E.S. & Hartley, H.O., 1966, *Biometrika Tables for Statisticians*, Vol. 1, 3rd edn. London : Cambridge University Press.

Pearson, K., 1900, On a criterion that a given system of deviations from the probable in the case of a correlated system of variables is such that it can be reasonably supposed to have arisen from random sampling. *Phil. Mag.* (5), **50**, 157–75.

Peers, H.W., 1971, Likelihood ratio and associated test criteria. *Biometrika,* **58**, 577–87.

Plackett, R.L., 1962, A note on interactions in contingency tables. *J.R. Statist. Soc. B*, **24**, 162–6.

Potthoff, R.F. & Whittinghill, M., 1966a, Testing for homogeneity: I. The binomial and multinomial distributions. *Biometrika*, **53**, 167–82.

Potthoff, R.F. & Whittinghill, M., 1966b, Testing for homogeneity. II. The Poisson distribution. *Biometrika*, **53**, 183–90.

Pringle, R.M. & Rayner, A.A., 1971, *Generalized Inverse Matrices with Applications to Statistics*. London : Griffin.

Quesenberry, C.P. & Hurst, D.C., 1964, Large sample simultaneous confidence intervals for multinomial proportions. *Technometrics*, **6**, 191–5.

Radhakrishna, S., 1965, Combination of results from several 2 × 2 contingency tables. *Biometrics*, **21**, 86–98.

Rao, C.R., 1961, Asymptotic efficiency and limiting information. *Proc. 4th Berkeley Symp. Math. Statist. Prob.*, **1**, 531–45.

Rao, C.R., 1965, *Linear Statistical Inference and its Applications.* New York : Wiley.

Rao, C.R. & Chakravarti, I.M., 1956, Some small sample tests of significance for a Poisson distribution. *Biometrics*, **12**, 264–82.

Rasch, G., 1960, *Probabilistic Models for some Intelligence and Attainment Tests.* Copenhagen: Danmarks Paedagogiske Institut: Nielsen & Lydiche.

Reiersøl, O., 1954, Tests of linear hypotheses concerning binomial experiments. *Skand. Akt.*, **37**, 38–59.

Ries, P.N. & Smith, H., 1963, The use of chi-square for preference testing in multidimensional problems. *Chemical Engineering Progress*, *Symposium Series*, **59**, (42), 39–43.

Roy, S.N. & Kastenbaum, M.A., 1956, On the hypothesis of no "interaction" in a multi-way contingency table. *Ann. Math. Statist.*, **27**, 749–57.

Scheffé, H., 1959, *The Analysis of Variance*. New York : Wiley.

Selby, B., 1965, The index of dispersion as a test statistic. *Biometrika*, **52**, 627–9.

Simpson, E.H., 1951, The interpretation of interaction in contingency tables. *J.R. Statist. Soc. B*, **13**, 238–41.

Sinkhorn, R., 1967, Diagonal equivalence to matrices with prescribed row and column sums. *Amer. Math. Mon.*, **74**, 402–5.

Skellam, J.G., 1952, Studies in statistical ecology: I. Spatial pattern. *Biometrika*, **39**, 346–62.

Slakter, M.J., 1966, Comparative validity of the chi-square and two modified chi-square goodness-of-fit tests for small but equal expected frequencies. *Biometrika*, **53**, 619–23.

Slakter, M.J., 1968, Accuracy of an approximation to the power of the chi-square goodness-of-fit test with small but equal expected frequencies. *J. Amer. Statist. Assoc.*, **63**, 912–24.

Snedecor, G.W. & Cochran, W.G., 1967, *Statistical Methods*, 6th edn. Iowa : State University Press.

Snell, E.J., 1964, A scaling procedure for ordered categorical data. *Biometrics*, **20**, 592–607.

Solomon, H., 1961, Classification procedures based on dichotomous response vectors. *Studies in Item Analysis and Prediction* (H. Solomon, ed.), pp.177–86. Stanford, Calif.: Stanford University Press.

Sowden, R.R., 1971, Bias and accuracy of parameter estimates in a quantal response model. *Biometrika*, **58**, 595–603.

Sowden, R.R., 1972, On the first-order bias of parameter estimates in a quantal response model under alternative estimation procedures. *Biometrika*, **59**, 573–9.

Stephens, M.A., 1969, A goodness-of-fit statistic for the circle, with some comparisons. *Biometrika*, **56**, 161–8.

Stevens, W.L., 1938, The distribution of entries in a contingency table with fixed marginal totals. *Ann. Eugen. Lond.*, **8**, 238–44.

Stuart, A., 1953, The estimation and comparison of strengths of association in contingency tables. *Biometrika*, **39**, 105–10.

Stuart, A., 1955, A test for homogeneity of the marginal distributions in a two-way classification. *Biometrika*, **42**, 412–16.

Thomas, D.G., 1971, Exact confidence limits for the odds ratio in a 2 × 2 table. *Appl. Statist.*, **20**, 105–10.

Tocher, K.D., 1950, Extension of the Neyman–Pearson theory of tests to discontinuous variates. *Biometrika*, **37**, 130–44.

Waite, H., 1915, Association of finger prints. *Biometrika*, **10**, 421–78.

Wald, A., 1943, *Sequential Analysis of Statistical Data : Theory*. Report submitted by the Statistical Research Group, Columbia University, to the Applied Mathematics Panel, National Defense Research Committee.

Wald, A., 1947, *Sequential Analysis*. New York : Wiley.

Walsh, J.E., 1963, Loss in test efficiency due to misclassification for 2 × 2 tables. *Biometrics*, **19**, 158–62.

Watson, G.S., 1956, Missing and "mixed-up" frequencies in contingency tables. *Biometrics*, **12**, 47–50.

Wedderburn, R.W.M., 1974, Generalised linear models specified in terms of constraints. *J.R. Statist. Soc. B*, **36**.

Weisberg, H.I., 1972, Bayesian comparison of two ordered multinomial populations. *Biometrics*, **28**, 859–67.

White, H.C., 1963, Cause and effect in social mobility tables. *Behavioral Science*, **8**, 14–27.

Williams, E.J., 1952, Use of scores for the analysis of association in contingency tables. *Biometrika*, **39**, 274–89.

Williams, O.D. & Grizzle, J.E., 1972, Analysis of contingency tables having ordered response categories. *J. Amer. Statist. Assoc.*, **67**, 55–63.

Winsor, C.P. 1948, Factorial analysis of a multiple dichotomy. *Human Biology*, **20**, 195–204.

Wisniewski, T.K.M., 1968, Testing for homogeneity of a binomial series. *Biometrika*, **55**, 426–8.

Woolf, B., 1955, On estimating the relation between blood group and disease. *Ann. Human Genet., Lond.*, **19**, 251–3.

Wynder, E.L. & Cornfield, J., 1953, Cancer of the lung in physicians. *New England J. Med.*, **248**, 441–4.

Yates, F., 1934, Contingency tables involving small numbers and the χ^2 test. *Suppl. J.R. Statist. Soc.*, **1**, 217–35.

Yates, F., 1948, The analysis of contingency tables with groupings based on quantitative characters. *Biometrika*, **35**, 176–81; corr., **35**, 424.

Yates, F., 1955a, The use of transformations and maximum likelihood in the analysis of quantal experiments involving two treatments. *Biometrika*, **42**, 382–403.

Yates, F., 1955b, A note on the application of the combination of probabilities test to a set of 2 × 2 tables. *Biometrika*, **42**, 404–11.

Yates, F., 1960, *Sampling Methods for Censuses and Surveys*, 3rd edn. London : Griffin.

Youkeles, Lee H., 1963, Loss of power through ineffective pairing of observations in small two-treatment all-or-none experiments. *Biometrics*, **19**, 175–80.

Zelen, M., 1971, The analysis of several 2 × 2 contingency tables. *Biometrika*, **58**, 129–37.

Zelen, M., 1972, Exact significance tests for contingency tables embedded in a 2^n classification. *Proc. 6th Berkeley Symp. Math. Statist. & Prob.* (ed. L.M. LeCam, J. Neyman & E.L. Scott), pp.737–57. Berkeley: University of California Press.

AUTHOR INDEX

SUBJECT INDEX

EXPERIMENT INDEX

Experimental units followed by methods of classification